Forensic and Environmental
Detection of Explosives

Forensic and Environmental Detection of Explosives

JEHUDA YINON

Department of Environmental Sciences and Energy Research,
Weizmann Institute of Science, Rehovot, Israel

JOHN WILEY & SONS, LTD
Chichester • New York • Weinheim • Brisbane • Singapore • Toronto

Other Wiley Editorial Offices

John Wiley & Sons, Inc., 605 Third Avenue,
New York, NY 10158-0012, USA

WILEY-VCH Verlag GmbH, Pappelallee 3,
D-69469 Weinheim, Germany

Jacaranda Wiley Ltd, 33 Park Road, Milton,
Queensland 4064, Australia

John Wiley & Sons (Asia) Pte Ltd, Clementi Loop #02-01,
Jin Xing Distripark, Singapore 129809

John Wiley & Sons (Canada) Ltd, 22 Worcester Road,
Rexdale, Ontario M9W IL1, Canada

Library of Congress Cataloging-in-Publication Data

Yinon, Jehuda.
    Forensic and environmental detection of explosives / Jehuda Yinon.
        p.   cm.
    Includes bibliographical references and index.
    ISBN 0-471-98371-3 (alk. paper)
    1. Explosives—Detection.   I. Title.
TP271.Y564   1999
363.25—dc21   99-13560
                                                                  CIP

British Library Cataloging in Publication Data

A catalogue record for this book is available from the British Library

ISBN 0 471 98371 3

Typeset in 10/12pt Times by Laser Words, Madras, India
Printed and bound in Great Britain by Antony Rowe Ltd, Chippenham, Wiltshire
This book is printed on acid-free paper responsibly manufactured from sustainable forestry,
in which at least two trees are planted for each one used for paper production.

To my wife Bracha, who displayed strength,
determination and dedication throughout
difficult times, and to whom I wish to
express my love and gratitude.

Contents

Preface

The detection of explosives is an extremely relevant analytical problem, which has become an issue of major importance in the forensic and environmental sciences.

Acts of terrorism, such as the destruction of Pan Am flight 103 over Scotland in 1988, the bombing of New York City's World Trade Center in 1993, and the Oklahoma City bombing in 1995, have brought terrorism to the awareness of the public in a tragic way. With the increasing use of explosives by terrorist groups and by individuals, law enforcement forces throughout the world are faced with the problem of detecting hidden explosives in luggage, mail, vehicles, aircraft, on suspects, etc., in order to prevent such acts of terrorism and crime.

There are two areas of endeavor requiring environmental detection of explosives. On-site detection and identification of traces of explosives and their degradation products, in areas suspected of being contaminated by toxic explosives, is necessary in order to monitor the quality of groundwater and prevent poisoning of populations of humans and animals. The detection of landmines is an acute and urgent worldwide problem which needs specific, rapid and cost-effective solutions. These are relative new areas of detection of explosives for which some specific techniques have already been developed.

Despite the differences between forensic and environmental applications, they have many common features. In both applications one has to deal with the detection and identification of very small amounts of explosives, and in both applications the lives of people could depend on the reliability of the detection methods. Some of the instrumental techniques are being used in both applications.

This book contains detailed descriptions of the principles and the methodology of the analytical instrumentation used in both forensic and environmental detection of explosives, as well as examples of the use of these instruments in both applications. Specific problems exist in each one of these applications and will be discussed. In forensic detection of explosives, problems include speed of detection, camouflage of explosive devices and matrix interference, i.e. the complexity of the composition of suspicious objects.

Although bombs can also be detected by recognition of their typical working parts, such as timers, switches, detonators, batteries, wires, etc., this book will deal only with the recognition of the explosive material.

In environmental detection of explosives, problems include the identification and characterization of degradation products of explosives and the variability of the matrix, which in most cases is soil or water.

A short overview and classification of the various explosives is followed by a glossary of basic terms to facilitate the reading of the subsequent chapters. Methodology of detection of concealed explosives has been divided in two main chapters: vapor detection (sniffers) and bulk detection methods. Because of its expanding application, a special chapter has been devoted to tagging of explosives, which has become in several countries a necessity by law. The chapter on environmental detection of explosives contains a short review on the toxicity of explosives and their degradation in soil and water, and specific methods for on-site detection of explosives in contaminated areas. Despite the major importance and broad range of mine detection methods, only a relatively short chapter has been included to describe this subject, due to the limited scope of the book. Although the book deals only with detection methods of explosives, this chapter also overviews briefly anomaly mine detectors, so as to form a whole entity.

The content of this monograph is based only on published material - from scientific journals, conference proceedings, unclassified reports and manufacturers' technical literature.

Finally, I would like to thank all those who have sent me material on the various subjects of this book. Many thanks are due to authors and publishers for permission to reproduce copyrighted figures.

Jehuda Yinon
Rehovot, May 1999

Abbreviations

Abbrev.	Definition
2,2′-Az	4,4′,6,6′-tetranitro-2.2′-azoxytoluene
4,4′-Az	2,2′,6,6′-tetranitro-4,4′-azoxytoluene
2,4-DA	2,4-diamino-6-nitrotoluene
2,6-DA	2,6-diamino-4-nitrotoluene
2-A	2-amino-4,6-dinitrotoluene
4-A	4-amino-2,6-dinitrotoluene
ADXRD	angular dispersive X-ray diffraction
AMP	adenosine monophosphate
AN	ammonium nitrate
ANFO	ammonium nitrate–fuel oil
AP	appearance potential
APCI	atmospheric pressure chemical ionization
API	associated particle imaging
API	atmospheric pressure ionization
ARC	accelerating rate calorimetry
ASGDI	atmospheric sampling glow discharge ionization
ATP	adenosine triphosphate
BDE	bond dissociation energy
BGO	bismuth germinate
BL	bioluminescence
BSA	bovine serum albumin
CCD	charge-coupled device
CFI	continuous flow immunosensor
CID	collision-induced dissociation
CL	chemiluminescence
CL-20	hexanitrohexaazaisowurtzitane
CT	computed tomography
CW	continuous wave
CXRS	coherent X-ray scattering

d.c.	direct current
DATB	1,3-diamino-2,4,6-trinitrobenzene
DDNP	diazodinitrophenol
DEGN	diethyleneglycoldinitrate
	dinitroglycerol
DINA	diethylnitramine dinitrate
DMNB	2,3-dimethyl-2,3-dinitrobutane
DNB	1,3-dinitrobenzene
DNT	2,4-dinitrotoluene
DNX	hexahydro-1,3-dinitroso-5-nitro-1,3,5-triazine
DOP	dioctyl phthalate
DSC	differential scanning calorimetry
EA	electron affinity
ECD	electron-capture detector
ECNIMS	electron-capture negative-ion mass spectrometry
EDA	ethylene diamine
EDD	explosive detection device
EDDN	ethylene diamine dinitrate
EDNA	ethylenedinitramine
	Haleite
EDS	explosive detection system
EGDN	ethylene glycol dinitrate
	glycoldinitrate
	nitroglycol
ELISA	enzyme-linked immunosorbent assay
EMI	electromagnetic induction
EPA	US Environmental Protection Agency
EPR	electron paramagnetic resonance
ESR	electron spin resonance
EXDT	energy-dispersive X-ray diffraction tomography
FMN	flavine mononucleotide
FMS	frequency modulation spectroscopy
FNA	fast neutron activation
FNTS	fast neutron transmission spectroscopy
FOPH	fiber-optic probe head
FT	fourier transform
GC	gas chromatography
GMBS	*N*-succinimidyl 4-maleimidobutyrate
GPR	ground penetrating radar
hexyl	2,2′,4,4′,6,6′-hexanitrodiphenylamine
	(dipicrylamine)
HFC	heat flux calorimetry
HMTD	hexamethylene triperoxide diamine

HMX	1,3,5,7-tetranitro-1,3,5,7-tetrazacyclooctane
	octahydro-1,3,5,7-tetranitro-1,3,5,7-tetrazocine
	cyclotetramethylenetetranitramine
	octogen
HNS	2,2′,4,4′,6,6′-hexanitrostilbene
HPLC	high performance liquid chromatography
HTNMR	hydrogen transient nuclear magnetic resonance
ICAO	International Civil Aviation Organization
IDP	isodecyl diphenylphosphate
IED	improvised explosive device
IMS	ion mobility spectrometry
IR	infrared
IT	ion trap
L-4412	perfluorohexylsulfur pentafluoride
LAXS	low-angle X-ray scattering
LC	liquid chromatography
LD_{50}	50% lethal dose value
LDL	lower detection limit
LIF	laser-induced fluorescence
LMR	lateral migration radiography
LOD	limit of detection
MATB	1-monoamino-2,4,6-trinitrobenzene
	picramide
	2,4,6-trinitroaniline
MEMS	microelectromechanical system
MHN	mannitol hexanitrate
	nitromannitol
MIR	micropower impulse radar
MMAN	monomethylamine nitrate
MMW	millimeter waves
MNX	hexahydro-1-nitroso-3,5-dinitro-1,3,5-triazine
MRM	multiple reaction monitoring
MS	mass spectrometry
MS/MS	tandem mass spectrometry
MTS	(3-mercaptopropyl)trimethoxysilane
NA	numerical aperture
NADH	reduced nicotinamide adenine dinucleotide
NC	nitrocellulose
Nd–YAG	neodymium–yttrium aluminum garnet laser
NEBS	neutron elastic back scatter
NES	neutron elastic scatter
NG	nitroglycerin
	glycerol trinitrate
	trinitroglycerol

NHS	N-hydroxysuccinimide
NMR	nuclear magnetic resonance
NQR	nuclear quadrupole resonance
NRA	nuclear resonance absorption
NRES	neutron resonant elastic scatter
o-MNT	o-mononitrotoluene
OVA	chicken egg albumin
p.p.b.	parts per billion ($\times 10^{-9}$ g/cm^3)
p.p.m.	parts per million ($\times 10^{-6}$ g/cm^3)
p.p.m.v.	parts per million by volume
p.p.t.	parts per trillion ($\times 10^{-12}$ g/cm^3)
PBA	pyrenebutyric acid
PBS	phosphate buffered saline
PDCB	perfluorodimethylcyclobutane
PDCH	perfluorodimethylcyclohexane
PET	positron emission tomography
PETN	pentaerythritol tetranitrate
	penta
PF	photofragmentation
PFD	perfluorodecalin
PFNA	pulsed fast neutron analysis
PMCH	perfluoromethylcyclohexane
p-MNT	p-mononitrotoluene
PPi	inorganic pyrophosphate
PVC	poly(vinyl chloride)
QRA	quadrupole resonance analysis
RDX	1.3.5-trinitro-1,3,5-triazacyclohexane
	hexahydro-1,3,5-trinitro-1,3,5-triazine
	cyclotrimethylenetrinitramine
	hexogen
READ	reversal electron attachment detection
RF	radio frequency
RGB	red–green–blue color model
RSD	relative standard deviation
RTS	real-time sampler
SAW	surface acoustic wave
SPME	solid phase microextraction
SRM	single reaction monitoring
SSFP	steady state free precession
STNG	sealed-tube neutron generator
TATB	1,3,5-triamino-2,4,6-trinitrobenzene

TATP	triacetone triperoxide
	acetonetriperoxide
	tricycloacetone peroxide
TAX	1-acetylhexahydro-3,5-dinitro-1,3,5-triazine
TEA	thermal energy analyzer (chemiluminescence detector)
TEGN	triethylene glycol dinitrate
	dinitrotriglycol
tetryl	2,4,6,N-tetranitro-N-methylaniline
	2,4,6-trinitrophenylmethylnitramine
TFR	thin-film resonator
TGA	thermal gravimetric analysis
THAM	tromethamine
TILDAS	tunable infrared laser differential absorption spectroscopy
TNA	thermal neutron activation
TNAZ	1,3,3-trinitroazetidine
TNB	trinitrobenzene
TNP-POD	trinitrophenyl-peroxidase conjugate
TNT	2,4,6-trinitrotoluene
TNX	hexahydro-1,3,5-trinitroso-1,3,5-triazine
TOF	time of flight
UXO	unexploded ordnance
XENIS	X-ray enhanced neutron interrogation system

1

Classification of explosives and basic terms

This chapter includes a classification of explosives which is based on the applications of existing explosives and on their contents, but also takes into account the capabilities and characteristics of the detection techniques. It includes a representative list of propellants, military, and industrial explosives. It does not contain a list of improvised explosive combinations, the number of which is unlimited and depends only on the ingenuity of the criminal or terrorist. Such a list was omitted from this book in order not to contribute to the irresponsible propagation of homemade bomb formulae.

This chapter contains general data on the basic explosives, mainly physical and chemical properties [1–19], but also health effects and information on environmental impact [1, 2, 11, 12, 14, 18, 20]. In addition, this chapter contains a glossary of basic terms related to the detection of explosives.

1.1 PROPERTIES OF BASIC EXPLOSIVES

1.1.1 2,4,6-Trinitrotoluene (TNT)

1.1.1.1 General properties

2,4,6-Trinitrotoluene (TNT) is one of the most widely used military high explosives. TNT is very stable, nonhygroscopic, and relatively insensitive to impact, friction, shock, and electrostatic energy.

Along with the 2,4,6-trinitrotoluene symmetrical isomer, five unsymmetrical isomers are found in the crude product, resulting from the nitration of toluene with nitric acid, in the presence of sulfuric acid. The nitration occurs in a stepwise mode by either a batch or a continuous process.

Wastewater from TNT manufacture is contaminated with organic nitrocompounds, nitrates, sulfates, and sodium sulfite and nitrite. Upon exposure to sunlight, some of these compounds undergo chemical transformation, producing highly colored substances ('pink water').

The addition of oxygen-rich products to TNT can form mixtures with enhanced explosive power. Various explosive compounds have been added to TNT to form binary explosives. Other compounds have been added in order to decrease sensitivity and increase mechanical strength of the cast.

Chemical and physical properties of TNT are presented in Table 1.1. The vapor pressure of TNT at various temperatures is shown in Table 1.2.

1.1.1.2 Health effects and environmental impact

TNT is absorbed by inhalation, ingestion, or skin contact. TNT poisoning can lead to severe diseases, such as aplastic anemia or toxic jaundice [2, 11]. In

Table 1.1 Chemical and physical properties of 2,4,6-trinitrotoluene (TNT) [1–4, 6–9]

Molecular weight	227.13
Molecular formula	$C_7H_5N_3O_6$
Structure	
Crystal form	Colorless orthorhombic crystals or yellow monoclinic needles.
Density (g/cm³)	1.65
Solubility characteristics (g/100 g)	
Water	0.01 (25°C)
	0.02 (42°C)
	0.14 (100°C)
Carbon disulfide	0.63 (25°C)
Carbon tetrachloride	0.82 (25°C)
Ethanol	1.5 (25°C)
Diethyl ether	3.8 (25°C)
Chloroform	25.0 (25°C)
Toluene	67.0 (25°C)
Benzene	88.0 (25°C)
Acetone	132.0 (25°C)
Melting point (°C)	80.65
Boiling point (°C)	210 (10 torr)
Flash point (°C)	240
Ignition point (°C)	295

Table 1.2 Vapor pressure of 2,4,6-trinitrotoluene (TNT)

Temperature (°C)	Vapor pressure (p.p.m.)	(torr)	Reference
25	7.7×10^{-3}	5.8×10^{-6}	[10]
85	70	0.053	[2]
100	140	0.106	[2, 4]
150	4×10^3	3.0	[4]
200	14×10^3	10.5	[4]

aplastic anemia (which is nearly always fatal) the blood-forming organs fail to function, resulting in a progressive loss of the blood elements. Toxic jaundice is an indication of severe liver damage. Less serious diseases resulting from TNT intoxication include dermatitis, gastritis, cyanosis, and various effects on the nervous system.

Disposal of TNT and its degradation products from munitions manufacturing plants and disposal sites presents a serious and potentially hazardous environmental problem. For many years, obsolete explosives and munitions were disposed of by burial in the ground and dumping in the sea. Wastewater from TNT manufacture and shell loading plants were disposed of by discharge into rivers or streams. As a result, large areas in many countries are contaminated by TNT, as well as by other explosives, and their degradation products.

1.1.2 1,3,5-Trinitro-1,3,5-triazacyclohexane (RDX)

1.1.2.1 General properties

RDX (cyclotrimethylenetrinitramine, hexahydro-1,3,5-trinitro-1,3,5-triazine) is an important military explosive. It has been used extensively as a booster charge in many munitions formulations, especially in artillery shells. RDX is relatively insensitive, it has a high chemical stability, although lower than that of TNT, and an explosive power greater than that of TNT.

RDX is manufactured by either one of two processes. In the Woolwich process, or direct nitrolysis, hexamine is reacted with nitric acid to produce RDX [3]. The product contains traces of HMX (see below). The Bachmann process involves nitration of hexamine with ammonium nitrate and nitric acid in an acetic acid–acetic anhydride solvent [13]. The product contains 6–9% of HMX.

RDX is never handled pure and dry, because of the danger of accidental explosion. Following production, it is immediately incorporated into formulations or desensitized with additive. RDX is used as a component in explosive mixtures, especially plastic explosives.

Table 1.3 Chemical and physical properties of RDX [1, 3, 6, 7, 9, 12]

Molecular weight	222.26
Molecular formula	$C_3H_6N_6O_6$
Structure	

Crystal form	White orthorhombic crystals.
Density (g/cm³)	1.83
Solubility characteristics (g/100 g)	
Water	0.006 (25°C)
Toluene	0.02 (20°C)
Benzene	0.05 (20°C)
Diethyl ether	0.055 (20°C)
Ethanol	0.11 (20°C)
Nitrobenzene	1.5 (25°C)
Methyl acetate	1.9 (20°C)
Acetic anhydride	4.9 (30°C)
Acetonitrile	5.5 (25°C)
Acetone	8.3 (25°C)
Cyclohexanone	12.7 (25°C)
Dimethyl formamide	37.0 (25°C)
Dimethyl sulfoxide	41.0 (25°C)
Melting point (°C)	204
Ignition point (°C)	295
Heat of combustion (kcal/g)	2.26

Chemical and physical properties of RDX are presented in Table 1.3. The vapor pressure of RDX at various temperatures is shown in Table 1.4.

1.1.2.2 Health effects and environmental impact

Symptoms and clinical manifestations of RDX poisoning include convulsions (epileptiform seizures) followed by loss of consciousness, muscular cramps, dizziness, headache, nausea, and vomiting [11, 12]. Absorption of RDX was reported to occur by inhalation or ingestion.

Potential environmental exposure to RDX exists not only in manufacturing plants where RDX dust is generated, but also in nearby waterways into which RDX might have been discharged and near RDX demilitarization

Table 1.4 Vapor pressure of RDX

Temperature (°C)	Vapor pressure (p.p.b.)	(torr)	Reference
25	6.0×10^{-3}	4.6×10^{-9}	[10]
43	0.1	7.6×10^{-8}	[10]
100	10^2	7.6×10^{-5}	[10]
200	1.2×10^5	0.09	[1, 6]

sites where obsolete explosives and ammunition containing RDX was buried in the ground.

1.1.3 1,3,5,7-Tetranitro-1,3,5,7-tetrazacyclooctane (HMX)

HMX (Octahydro-1,3,5,7-tetranitro-1,3,5,7-tetrazocine, cyclotetramethylenetetranitramine, octogen) has been used as a component in solid-fuel rocket propellants and as burster charges for artillery shells.

HMX is manufactured by a process similar to the Bachmann process, used for the manufacture of RDX, with some modifications [6]. HMX and RDX have similar chemical reactivities [3]. The only difference, which is used to separate HMX from RDX, is that HMX is more resistant than RDX to the action of sodium hydroxide.

Chemical and physical properties of HMX are shown in Table 1.5.

Very little is known about the human toxicity of HMX. Clinical signs of acute toxicity observed in mice and rabbits, fed with doses of HMX, included clonic convulsions and spasms, ataxia, and other clinical effects related to the central nervous system [14].

As for RDX, potential environmental exposure to HMX exists as a result of discharge of HMX-contaminated wastewater into rivers and streams and dumping of obsolete explosives and munitions into landfill sites.

1.1.4 Tetryl

Tetryl (2,4,6,*N*-tetranitro-*N*-methylaniline, 2,4,6-trinitrophenylmethylnitramine) has been used as a booster explosive and as a base charge in detonators and blasting caps [6].

Tetryl is much more sensitive to impact and friction than TNT. It has a greater rate of detonation and explosive power than TNT. It is only slightly hygroscopic and very stable. Although the production and use of tetryl has mostly been discontinued, it still can be found in some formulations and in some landmines.

The purified compound has a colorless crystalline form immediately after preparation, but it rapidly acquires a yellow color when exposed to light [4].

Table 1.5 Chemical and physical properties of HMX [1, 6, 7, 9, 14, 15]

Molecular weight	296.16
Molecular formula	$C_4H_8N_8O_8$
Structure	

Crystal form	Colorless crystalline solid. Four polymorphic forms: orthorhombic (α), monoclinic (β), monoclinic (γ) and hexagonal (δ). The β form is the least sensitive to impact and the most stable.
Density (g/cm³)	1.96 (β form)
Solubility characteristics (g/100 g)	
Water	0.00066 (20°C)
	0.00116 (30°C)
	0.00174 (35°C)
	0.014 (83°C)
Acetonitrile	2.0 (25°C)
Dimethyl formamide	2.3 (25°C)
Acetone	2.8 (25°C)
Dimethyl sulfoxide	57.0 (25°C)
Melting point (°C)	276–280
Ignition point (°C)	335
Heat of combustion (kcal/g)	2.25–2.36
Vapor pressure (at 100°C)	3.95 p.p.t. (3×10^{-9} torr)

The technical product has a pale lemon or buff color. When crystallized from benzene, it forms monoclinic prisms.

Chemical and physical properties of tetryl are presented in Table 1.6.

Tetryl is produced by a batch process involving sulfation of dimethylaniline with sulfuric acid to produce dimethylaniline sulfate, followed by nitration with a mixture of sulfuric and nitric acid, first to 2,4-dinitro-dimethylaniline and then to tetryl [6].

The most common symptom of exposure to tetryl is dermatitis [11]. Other symptoms include gastric disorders, headache, fatigue, and staining of the skin of the hands, face, and hair with a yellow color.

Table 1.6 Chemical and physical properties of tetryl [3, 4, 6, 9, 11]

Molecular weight	287.15
Molecular formula	$C_7H_5N_5O_8$
Structure	

Crystal form	monoclinic yellow crystals
Crystal density (g/cm^3)	1.73
Solubility characteristics (g/100 g)	
Water	0.008 (25°C)
Carbon tetrachloride	0.031 (25°C)
Diethyl ether	0.46 (25°C)
Ethanol	0.65 (25°C)
Toluene	0.25 (19.5°C)
Benzene	3.00 (20°C)
Ethyl acetate	12.2 (18°C)
Acetone	68.0 (25°C)
Dimethyl formamide	114.0 (25°C)
Melting point (°C)	129.5
Heat of combustion (kcal/g)	2.92
Vapor pressure (at 25°C) (torr)	5.7×10^{-9}

1.1.5 Pentaerythritol tetranitrate (PETN)

PETN (Penta) has been used as a base charge in blasting caps and detonators, as the core explosive in commercial detonating cord, in booster charges, and in plastic explosives. PETN is manufactured by nitration of pentaerythritol by a batch process at 15–25°C with concentrated nitric acid. The yield of the process is about 95%. Impurities present in the final product include pentaerythritol trinitrate, dipentaerythritol hexanitrate, and tripentaerythritol acetonitrate [6].

The chemical stability of PETN is very high [3]. PETN undergoes, quite rapidly, hydrolysis in water or in dilute nitric acid solutions at temperatures in the range 90–125°C. Because of its high sensitivity to impact, PETN is generally used with phlegmatizing additives or in mixtures. PETN is not very sensitive to friction, but is very sensitive to initiation by explosion. It is nonhygroscopic.

Chemical and physical properties of PETN are presented in Table 1.7.

Table 1.7 Chemical and physical properties of PETN [3, 6, 9, 10, 16]

Molecular weight	316.2
Molecular formula	$C_5H_8N_4O_{12}$
Structure	

$$O_2NOCH_2\text{---}\overset{\displaystyle CH_2ONO_2}{\underset{\displaystyle CH_2ONO_2}{\overset{\displaystyle |}{\underset{\displaystyle |}{C}}}}\text{---}CH_2ONO_2$$

Crystal form	White tetragonal crystals
Density [g/cm^3]	1.78
Solubility characteristics [g/100 g]	
Water	0.0002 (20°C)
Chloroform	0.06 (19°C)
Ethanol	0.2 (20°C)
Toluene	0.23 (20°C)
Diethyl ether	0.25 (20°C)
Benzene	0.30 (20°C)
Ethyl acetate	6.3 (19°C)
Acetone	20.3 (20°C)
Melting point (°C)	141.3
Heat of combustion (kcal/g)	1.96
Vapor pressure	
at 25°C	18 p.p.t. (1.4×10^{-8} torr)
at 97°C	1.1 p.p.m. (8.38×10^{-4} torr)
at 139°C	93 p.p.m. (7.08×10^{-2} torr)

PETN is absorbed by inhalation and by ingestion. Large doses of PETN have been shown to cause changes in the respiration, a fall of the blood pressure, a primary moderate rise in the venous pressure, and a considerable rise in the spinal pressure, lasting for several hours [7]. The clinical effects of PETN are considerably less than those of nitroglycerin (see paragraph 1.1.6).

Potential environmental exposure to PETN exists as a result of discharge of PETN-contaminated wastewater into rivers and streams and dumping of obsolete explosives and munitions into landfill sites.

1.1.6 Nitroglycerin (NG)

1.1.6.1 General properties

Nitroglycerin (NG, glycerol trinitrate, trinitroglycerol) is a colorless, odorless viscous liquid, with a burning sweetish taste. The commercial product has a pale yellow color. NG has been the main component in many dynamites and an ingredient in multibase propellants [3, 6]. NG is very sensitive to shock, impact and friction, and is used only when desensitized with other liquids or absorbent solids, or mixed with nitrocellulose. The sensitivity of NG decreases with decreasing temperature. Solid NG is less sensitive to shock than the liquid, although it is more sensitive to friction, because of intercrystalline contact [6]. Liquid NG starts to decompose at 50–60°C. At 145°C, the rate of decomposition is so rapid that the liquid appears to boil. NG is prepared by the Biazzi process, consisting of nitration of glycerin with a mixture of concentrated nitric and sulfuric acids [13]. Chemical and physical properties of NG are presented in Table 1.8.

Table 1.8 Chemical and physical properties of nitroglycerin (NG) [1, 6, 7, 9, 10, 18]

Molecular weight	227.09
Molecular formula	$C_3H_5N_3O_9$
Structure	CH_2—ONO_2
	CH—ONO_2
	CH_2—ONO_2
Crystal form	Rhombic
Density (g/cm^3)	1.59 (at 20°C)
Solubility characteristics [g/100 g at 20°C]	
Water	0.15
Carbon disulfide	1.25
Carbon tetrachloride	2.0
Ethanol	54.0
Diethyl ether	∞
Chloroform	∞
Benzene	∞
Acetone	∞
Ethyl acetate	∞
Melting point (°C)	13.2
Hygroscopicity (at 20°C)	0.06% at 90% relative humidity
Ignition point (°C)	250–260
Heat of combustion (kcal/g)	1.58
Vapor pressure	
at 20°C	0.34 p.p.m. (2.6×10^{-4} torr)
at 26°C	0.41 p.p.m. (3.1×10^{-4} torr)
at 93°C	400.0 p.p.m. (0.31 torr)

1.1.6.2 Health effects and environmental impact

NG can be absorbed by inhalation, ingestion, or absorption through the skin. Symptoms and clinical manifestations due to NG poisoning are mainly caused by reduction of blood pressure [11]. They include headache, throbbing in the head, palpitation of the heart, nausea, vomiting, and flushing. In severe cases, the heart muscle is affected directly, with typical symptoms, such as weakening of the heart beats, delirium, convulsions, and even paralysis of the respiratory center. After continuous exposure to NG, an interruption of the exposure may be followed by pains in the chest. These symptoms are called 'withdrawal symptoms' and may sometimes result in death [20].

Potential exposure to NG exists in manufacturing plants of NG and NG-containing munitions. It is assumed that, because of its solubility in water, NG, contained in discharged wastewater, will be carried by streams without settling in reservoirs or lakes [11]. However, NG contained in munitions buried in the ground, might find its way into groundwater.

1.1.7 Ethylene glycol dinitrate (EGDN)

Ethylene glycol dinitrate (EGDN, nitroglycol, glycol dinitrate) is a transparent, colorless, liquid explosive. EGDN is a good solvent for low-grade nitrocellulose, it is comparable to NG in explosive energy, but is less sensitive to impact and more stable [1, 6]. EGDN is made by nitration of ethylene glycol with mixed acid, at a yield of about 93%.

EGDN has been used in mixtures with NG for low-temperature dynamites. However, the increasing use of ammonium nitrate–fuel oil (ANFO) and slurry explosives to replace dynamites, has greatly decreased the use of EGDN.

Chemical and physical properties of EGDN are presented in Table 1.9.

The toxic and environmental effects of EGDN are similar to those of NG. However, because of the higher vapor pressure of EGDN, exposure to EGDN results in more acute symptoms and clinical manifestations.

1.1.8 Triacetone triperoxide (TATP)

TATP (acetonetriperoxide, tricycloacetone peroxide) has been recommended for use in primers and detonators, but owing to its high volatility and high sensitivity, it has not been applied for military use [19]. It has, however, been used in recent years in terrorist activities.

TATP is prepared by mixing equal amounts of acetone and 50% hydrogen peroxide, with addition of hydrochloric acid, phosphoric acid, or sulfuric acid.

TATP explodes violently upon heating, impact or friction. It is highly brisant and very sensitive to friction. Some chemical and physical properties of TATP are shown in Table 1.10.

Equal weights of TATP and other explosives, stored for 40 days at 50°C, demonstrated a weight loss of about 50%, owing to complete volatilization of the TATP. There was no decomposition of the other explosives.

Table 1.9 Chemical and physical properties of ethylene glycol dinitrate (EGDN) [1, 6, 9]

Molecular weight	152.1
Molecular formula	$C_2H_4N_2O_6$
Structure	$CH_2 \!\!-\!\! ONO_2$
	$\mid$
	$CH_2 \!\!-\!\! ONO_2$
Density [g/cm^3]	1.49
Solubility characteristics [g/100 g at 20°C]	
Water	0.15
Diethyl ether	∞
Chloroform	∞
Benzene	∞
Toluene	∞
Acetone	∞
Melting point (°C)	−22.8
Hygroscopicity	Nonhygroscopic
Heat of combustion (kcal/g)	1.76
Vapor pressure	
at 25°C	92.6 p.p.m. (0.07 torr)
at 35°C	290 p.p.m. (0.22 torr)
at 55°C	1260 p.p.m. (0.96 torr)

Results of impact sensitivity tests show that TATP is one of the most sensitive explosives known. The explosive power of TATP is comparable to that of TNT.

1.1.9 Ammonium nitrate (AN)

Ammonium nitrate (AN) is an odorless white compound which is used as a solid oxidizer in explosive mixtures. AN is also widely used as a fertilizer. AN has a good chemical stability and a low sensitivity to friction and shock [6]. It is very hygroscopic and deliquesces at a relative humidity of about 60%. AN exists in five crystal forms which may transform from one to the other with accompanying volume, crystal structure, and heat changes.

AN is manufactured from anhydrous ammonia and nitric acid, which is itself obtained by oxidation of ammonia [1, 19]. It ranges from dense crystals to porous prills. Prills used in industrial explosives are made by spraying a 95% solution of the nitrate against a countercurrent stream of air. The particles are dried and coated to improve flow characteristics and moisture resistance. Pure

Table 1.10 Chemical and physical properties of triacetone triperoxide (TATP) [19]

Molecular weight	222.23
Molecular formula	$C_9H_{18}O_6$
Structure	

Density (g/cm^3)	1.2

Solubility characteristics (g/100 g) at 17°C

Ethanol	0.15
Diethyl ether	5.5
Petrol ether	7.35
Acetone	9.15
Carbon disulfide	9.97
Pyridine	15.4
Benzene	18.0
Trichloroethylene	22.7
Carbon tetrachloride	24.8
Chloroform	42.5

Melting point (°C)	94–95

Volatility

at 14–18°C	sublimates, losing about 6.5% of its weight in 24 h
at 25°C	loses 68% of its weight in 14 days
at 50°C	loses 1.5% of its weight in 2 h
at 100°C	volatilizes very rapidly, depositing fine needles

AN, when hermetically confined, can be made to explode by rapid heating to a temperature above 200°C.

The manufacture of fertilizer-grade AN differs from explosive-grade AN only in that the sprayed solution is a 99% AN concentration, rather than the 95% concentration used for explosive-grade AN. The resulting prill, nearly a solid sphere of AN with little porosity, is still a detonable material when mixed with a fuel, although it requires a larger booster and cannot be used in small-diameter boreholes.

No toxic effects of AN have been reported in the literature.

Chemical and physical properties of AN are presented in Table 1.11.

Table 1.11 Chemical and physical properties of ammonium nitrate (AN) [1, 3, 6, 7, 10, 19]

Molecular weight	80.05
Molecular formula	NH_4NO_3
Crystal form	Five crystal forms: orthorhombic at room temperature
Density (g/cm^3)	1.59
Solubility characteristics (g/100 g)	
Water	66.1 (20°C)
Methanol	14.0 (18.5°C)
Ethanol	2.5 (20°C)
Melting point (°C)	169.6
Heat of combustion (kcal/g)	0.626
Vapor pressure	
at 25°C	12 p.p.b. (9.1×10^{-6} torr)
at 35°C	29 p.p.b. (22.0×10^{-6} torr)
at 50°C	103 p.p.b. (78.3×10^{-6} torr)
at 80°C	870 p.p.b. (661.0×10^{-6} torr)

1.1.10 Various explosives

ammonium picrate (Explosive D)

2,4,6-trinitrophenol (picric acid)

2,4,6-trinitroaniline (picramide, 1-monoamino-2,4,6-trinitrobenzene, MATB)

1,3-diamino-2,4,6-trinitrobenzene (DATB)

1,3,5-triamino-2,4,6-trinitrobenzene (TATB)

2,2′,4,4′,6,6′-hexanitrostilbene (HNS)

2,2′,4,4′,6,6′-hexanitrodiphenylamine (hexyl, dipicrylamine)

diethylene glycol dinitrate (DEGN, dinitroglycol)

triethylene glycol dinitrate (TEGN, dinitrotriglycol)

nitrocellulose (NC)
structural formula (as the trinitrate)

mannitol hexanitrate (MHN, nitromannitol)

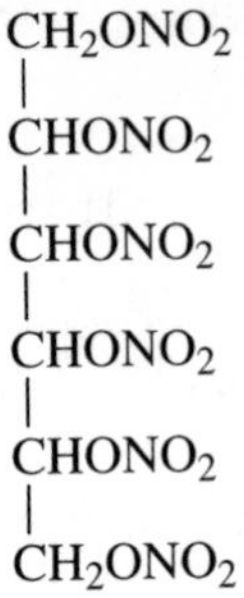

$$CH_2ONO_2$$
$$|$$
$$CHONO_2$$
$$|$$
$$CHONO_2$$
$$|$$
$$CHONO_2$$
$$|$$
$$CHONO_2$$
$$|$$
$$CH_2ONO_2$$

ethylene diamine dinitrate (EDDN)

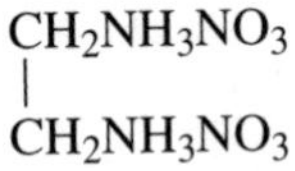

$$CH_2NH_3NO_3$$
$$|$$
$$CH_2NH_3NO_3$$

monoethanolamine dinitrate

$$NO_3NH_3CH_2CH_2ONO_2$$

monomethylamine nitrate (MMAN)

$$CH_3NH_3NO_3$$

ethylenedinitramine (EDNA, Haleite)

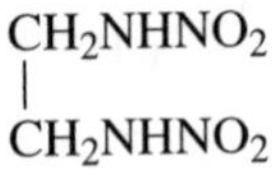

$$CH_2NHNO_2$$
$$|$$
$$CH_2NHNO_2$$

diethylnitramine dinitrate (DINA)

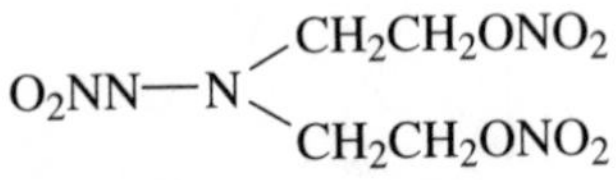

$$O_2NN-N\begin{array}{l} {}^{\diagup}CH_2CH_2ONO_2 \\ {}_{\diagdown}CH_2CH_2ONO_2 \end{array}$$

nitroguanidine

$$HN{=}C\begin{array}{l} {}^{\diagup}NH_2 \\ {}_{\diagdown}NH \\ \quad | \\ \quad NO_2 \end{array}$$

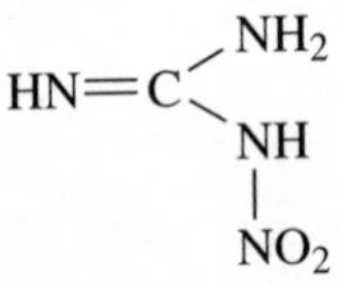

dinitroglycoluril (Dingu)

$$O_2N-N \begin{matrix} & H & H \\ & | & | \\ N-C-N \end{matrix}$$

O=C, N-C-N, C=O with H, H, NO$_2$ substituents

tetranitroglycoluril (Sorguyl)

hexamethylene triperoxide diamine (HMTD)

$$N \begin{matrix} CH_2-O-O-CH_2 \\ CH_2-O-O-CH_2 \\ CH_2-O-O-CH_2 \end{matrix} N$$

1,3,3-trinitroazetidine (TNAZ)

hexanitrohexaazaisowurtzitane (CL-20)

1.2 CLASSIFICATION OF EXPLOSIVES

1.2.1 Military explosives

Following is a representative list of military explosives [1, 19]. The percentages of the ingredients of each compound vary between manufacturers.

Amatol: TNT + AN
Ammonal (Minol): TNT + AN + Al
Composition A-3: RDX + wax
Composition B: RDX + TNT + wax
Composition C-4: RDX + polyisobutylene + di(2-ethylhexyl)sebacate + fuel oil
Cyclotol: RDX + TNT
DBX: TNT + RDX + AN + Al
Detasheet: PETN + plasticizer
H-6: RDX + TNT + Al + wax + calcium chloride
HBX-1: RDX + TNT + Al + wax + calcium chloride
Hexal: RDX + Al + wax
LX-10: HMX + Viton (fluorocarbon binder)
LX-17: TATB + Kel-F 800 (chlorofluorocarbon binder)
Octol: HMX + TNT
PBX-9404: HMX + NC + chloroethylphosphate
PBXN-107: RDX + plasticizer
PE 4: Same composition as C-4
Pentolite: PETN + TNT
Picratol: TNT + ammonium picrate
PTX-1: RDX + TNT + tetryl
PTX-2: RDX + TNT + PETN
Semtex-H: RDX + PETN + poly(butadiene–styrene) + oil
Tetrytol: TNT + tetryl
Torpex: TNT + RDX + Al
Trigonol: TNT + Al
Tritonal: TNT + Al

1.2.2 Nitroglycerin-based dynamites

Nobel's original dynamite contained about 75% NG absorbed into 25% diatomaceous earth (Kieselguhr), which served as desensitizer. The Kieselguhr has been replaced with a variety of other substances, including oxidizers, which increase the energy of the dynamites [1, 6].
Examples of some dynamites are:

Dynamite: NG + EGDN
Dynamite: NG + AN + NC + 2,4-dinitrotoluene (DNT)
Dynamite 3: NG + NC + sodium nitrate

Gomme A: NG + EGDN + NC
Wetter-Carbonit C: NG + EGDN + inorganic salts

1.2.3 Ammonium nitrate-based explosives [1, 6]

Ammonit 2: AN + TNT + DNT + woodmeal
Ammonium nitrate–fuel oil (ANFO) explosives:
  Lambrit: AN + fuel oil
  Prillex: AN + diesel fuel
Cava 1a: AN + TNT + wood meal
Cava 1n: AN + NG + EGDN + DNT + NC + woodmeal
Donarit 1: AN + EGDN + TNT + DNT + woodmeal
Frangex: AN + NG + DNT + NC + woodmeal
Gelamon 30: AN + EGDN + DNT + woodmeal
Gelamon 40: AN + EGDN + DNT + NC + sodium nitrate
Gelsurite 2000: AN + sodium nitrate + calcium nitrate + Al
Knauerit 2: AN + EGDN + TNT + DNT + NC + PETN + RDX + woodmeal
Magnafrac: AN + sodium nitrate + Al
Minol 2: AN + TNT + Al
Powermax 140: AN + sodium perchlorate + sodium nitrate
Titanite: AN + TNT

1.2.4 Slurry and emulsion explosives [1, 21]

1.2.4.1 Gel or slurry explosive

These explosives consist of a water solution of an inorganic oxidizer, such as AN, with mixtures of ammonium, sodium or calcium nitrate, gelled with a natural polysaccharide, such as guar gum. Other ingredients may be added such as fuels (Al), sensitizers (nitro explosives, organic amine nitrates), stabilizers, etc. Microballoons are added to adjust the sensitivity and density of the water gel.

  Some examples of gel explosives are the following.

CS Booster: AN + ethylene glycol mononitrate + gum + microspheres
Explogel: AN + MMAN + gelling agent and cross-linker + Al + perlite
Slurmex 200: AN + sodium nitrate + ethylene glycol mononitrate + Al + calcium nitrate + density controllers
Tovex: AN + MMAN + Al + starches + rubber + sodium nitrate + potassium pyroantimonate
Tutagex 110: AN + sodium nitrate + MMAN + guar gum + gilsonite + perlite

1.2.4.2 Emulsion explosives

Emulsions have, in addition to the ingredients listed in gel explosives, an oil/wax and emulsifier mixture. This is used in place of the gelling agent, to hold the ingredients together and provide structural integrity. A typical emulsion consists of water, one or more inorganic nitrate oxidizers, oil (with or without dissolved wax) and emulsifying agents of the water-in-oil type. Emulsions may contain chemical sensitizers, such as metal perchlorates, to improve initiation at low temperatures, but often rely on air bubbles or microspheres, for sensitization as well as density control.

Examples of emulsion explosives are as follows.

Emex: AN + sodium nitrate + fuel oil + wax and emulsifiers, sensitized with plastic microspheres

Emulex 720: AN + sodium nitrate + oil + Al + emulsifier + microballoons

ICI 7D: AN + ethylenediamine dinitrate + ammonium perchlorate + oil+wax + Al + emulsifier + microballoons

Lambrex 1: AN + sodium nitrate + fuel oil + wax + emulsifier

Lawinit 2: AN + sodium nitrate + fuel oil + wax + emulsifier

1.2.5 Low-order explosives (propellants)

Low-order explosives or propellants are used to propel projectiles from guns, to propel rockets and missiles and launch torpedoes. Low explosives are mostly solid combustible materials that decompose rapidly (deflagration), but do not normally explode [1, 6].

They contain one or more energetic materials, plasticizers to improve processing characteristics, stabilizers to increase storage life, and inorganic additives to facilitate handling, improve ignitability, and decrease muzzle flash.

1.2.5.1 Black powder

Black powder is composed essentially of a mixture of potassium nitrate or sodium nitrate, charcoal, and sulfur. It is hygroscopic and subject to rapid deterioration exposed to moisture. It is also one of the most dangerous explosives to handle because of the ease with which it is ignited by heat, friction or spark.

Although the use of black powder as a propellant and as a projectile bursting charge has ceased, it remains an explosive that is bound to be encountered in ammunition and ammunition components.

An example of a black powder is Pyrodex, which contains potassium nitrate, potassium perchlorate, charcoal, sulfur, cyanoguanidine, sodium benzoate, and dextrin.

1.2.5.2 Smokeless powder

Smokeless powder is used almost exclusively as the propellant for gun and rocket ammunition. During manufacture it is grained to a uniform size in the form of flakes, strips, sheets, balls, cords, or perforated cylindrical grains. Smokeless powder propellants are classified as single-base, double-base, and triple-base propellants. Single-base propellants include compositions that are principally gelatinized nitrocellulose and contain no high-explosive ingredient such as NG. A representative composition includes nitrocellulose, diphenylamine, DNT, and dibutyl phthalate. Double-base propellants are mainly compositions that include predominantly nitrocellulose and NG. An example is Ballistite, a rocket propellant, which is composed of nitrocellulose and NG, blended with diphenylamine, which acts as a stabilizer.

A triple-base propellant includes an additional explosive component. For example, Cordite N, which is used as a propellant in aircraft gun ammunition, contains nitrocellulose, NG and nitroguanidine. Composite propellants are compositions that contain mixtures of fuel and inorganic oxidants, but do not contain a significant amount of nitrocellulose or NG.

1.2.6 Special purpose explosives (detonating cords, blasting caps, and primers)

1.2.6.1 Initiating explosives (primers)

Initiating explosives or primers are highly sensitive to shock, friction, and heat, and are readily ignited by direct contact with flame or electrical sparks [1, 3, 6].

Their released energy and detonation velocity are small, but sufficient to detonate high explosives by shock wave. They are used in military detonators and in industrial blasting caps.

1.2.6.1.1 Mercury fulminate
Mercury fulminate ($Hg(ONC)_2$) is a heavy, practically nonhygroscopic, crystalline solid. When dry, it is very sensitive to heat, friction, spark, flame, and shock. It reacts with metals, such as aluminum, magnesium, zinc, brass, or bronze, in a moist atmosphere. Being a mercury derivative, it is a toxic compound.

1.2.6.1.2 Lead azide
Lead azide ($Pb(N_3)_2$) has a high temperature of ignition and is less sensitive to shock and friction than mercury fulminate. Because it is not easily ignited, lead azide is used in priming mixtures with lead styphnate (see below), which is very easy to ignite. Contact with copper must be avoided because it leads to the formation of the extremely sensitive copper azide.

1.2.6.1.3 Lead styphnate
Lead styphnate ($C_6HN_3O_8Pb \cdot H_2O$) is thermally stable, noncorrosive and non-hygroscopic. It is particularly sensitive to fire and to the discharge of static electricity. When dry, it can be readily detonated by static discharges from the human body. It does not react with metals, and is less sensitive to shock and friction than mercury fulminate or lead azide. It is used as a component in primer mixtures.

1.2.6.1.4 Tetrazene (tetracene)

Tetrazene ($C_2H_6N_{10} \cdot H_2O$) is only slightly hygroscopic and explodes readily from flame, producing a large amount of black smoke. It decomposes in boiling water. It ignites readily and is slightly more sensitive to impact than mercury fulminate. Its main use is for the sensitization of priming compositions.

1.2.6.1.5 Diazodinitrophenol (DDNP)

Diazodinitrophenol ($C_6H_2N_4O_5$) is less sensitive to impact, but more powerful than mercury fulminate and lead azide. It is nonhygroscopic and sparingly soluble in water. It is much less sensitive to friction than mercury fulminate but similar to that of lead azide. It is used as initiator in priming mixtures.

1.2.6.2 Detonating cords

Detonating cords usually consist of finely powdered PETN contained in a seamless cotton tube, with consecutive asphalt, rayon, and plastic cover layers. The amount of PETN is 1–40 g/m, mostly 10 g/m [22].

1.2.6.3 Blasting caps

Blasting caps consist of a base charge, an intermediate charge and ignition charge. Electric blasting caps are initiated by an electric spark. Nonelectric blasting caps are initiated by a safety fuse, which consists of a black powder core, a fiber wrap, and a waterproof cover.

1.3 BASIC TERMS

Absorbent A material able to absorb explosive vapors in order to concentrate them. Examples of such materials are platinum and Tenax. After concentration the explosive vapor is desorbed, usually by heat, into the detector.

Adsorption Gathering of a gas or liquid on a surface in a condensed layer (i.e. adsorption of a gas on charcoal). Explosive vapor molecules, on their way from the sampling device to the detector, are bound to be adsorbed to metal, glass, or plastic walls and surfaces. This will make them unavailable for detection and might cause a memory effect in subsequent detections.

Algorithm A step-by-step procedure for solving a problem, especially by a computer.

Antipersonnel mine An explosive, normally encased, designed to wound, kill or otherwise incapacitate personnel. It may be detonated by the action of its victim, by the passage of time or by controlled means.

Antitank mine A mine which is designed to disable or destroy vehicles and tanks. The explosive can be activated by pressure, tilt rod or command detonated.

Blasting agent Any material or mixture consisting of fuel and oxidizer intended for blasting, not otherwise defined as an explosive.

Bomblet A term used to describe types of submunitions, especially those packed within cluster bombs. Bomblets are designed to explode on contact with the target or ground.

Booster An explosive charge, usually a high explosive, used to initiate a less sensitive explosive.

Booby trap A device which is designed to explode, and which functions unexpectedly when a person or object (vehicle) disturbs or approaches an apparently harmless object.

Brisance The rapidity with which an explosive develops its maximum pressure; shattering effect.

Bulk detection Detection of explosives, using a radiation method.

Cartridge A cartridge containing an absorbent, which is part of a portable sampler. After sampling, the cartridge is removed from the sampler and introduced into the detection system.

Collection efficiency (of absorbing material) The efficiency of collecting explosives vapor on an absorbing material.

Concentrator A device containing an absorbent for concentration of explosive vapor.

Countermeasures Measures taken to circumvent a detection system.

Desorption Desorption, usually by heat, of concentrated explosive vapor into the detector.

Detection rate The rate at which pieces of luggage are passed through the detector. This rate should be as high as possible, but the inspection time should be long enough to allow complete examination of the item.

Detonating cord A flexible cord containing a center cord of high explosive that may be used to initiate other high explosives.

Detonator A device containing an initiating or primary explosive used to initiate detonation in another explosive.

Disarming The act of making a mine safe by removing the fuse or igniter. The procedure normally removes one or more links from the firing chain.

EDD Explosive detection device. An instrument which incorporates a single detection method to exploit one or more common physical property of explosives for detection purposes.

EDS Explosive detection system. This is a self-contained unit, composed of one or more devices (EDDs) integrated into a system.

False alarm Alarm given by the detector to signal the detection of an explosive in an examined item, although no explosive is present.

False alarm rate How often innocent objects will appear as explosives.

False negative The failure to detect an explosive in a tested item.

False positive The explosive detector will give an alarm to signal the detection of an explosive in a tested item, although no explosive is present.

Fingerprint Characteristic features (i.e. mass spectrum) of a compound.

Fraction of detection, $f(d)$ Number of detections divided by total number of possible detections.

Fraction of false alarms, $f(f_a)$ Number of false alarms divided by the total number of possible false alarms.

High explosive An explosive which is characterized by a very high rate of reaction, high pressure development and formation of a detonation wave, and which can be detonated by means of a blasting cap when unconfined.

High-Z material A material which has a high atomic number.

Improvised explosive device (IED) An improvised bomb, of local manufacture, which has the elements of a mine or a booby trap.

Inspection time The minimal time necessary for a complete inspection of a piece of luggage, a parcel, or a person, by a detector.

Interferences Materials or objects present in an examined item which could interfere with the detection procedure.

Interferrant A substance that causes a response in a particular detection instrument that is similar enough to a real explosive that its response cannot be discriminated from a real explosive.

Interrogation technique A detection technique for inspection of objects for hidden explosives.

Interrogated object An object inspected for hidden explosives.

Letter bomb A booby-trapped letter which contains a bomb.

Low explosive An explosive which is characterized by deflagration (a rapid combustion that moves through an explosive material at a velocity less than the speed of sound).

Lower detection limit (LDL) Smallest amount of explosive which is possible to detect.

Marked explosive An explosive, marked (tagged) during production, with an additive (tag, taggant) to make detection or identification easier. There are detection and identification tags.

Microparticle Trace of solid residue of an explosive.

Multisensor detection Detection system which includes more than one detecting sensor, not necessarily of the same type.

Nonintrusive Without forcing in. Nonintrusive detection of explosives means that it is not necessary to open a container or a piece of luggage, in order to search for hidden explosives. It can be done by radiation or by sniffing.

Olfaction Sense of smell (a dog capable of detecting explosives).

Olfactory Relating to the sense of smell.

Particle detection Detection of particles of explosives left on the surface of the container holding the explosives.

Penetrating radiation Radiation going through a checked item in order to detect hidden explosives.

Pixel Any of the small discrete elements that together constitute an image.

Portal Screening portal.

Potential threat Possibility of a hidden explosive in a tested item.

Preconcentration Concentration of explosive vapor before submitting to the detector.

Probing radiation Radiation passing through a tested item in order to check whether there are hidden explosives.

Prodder (or probe) A tool, consisting of one or more pointed rods or tines, used to probe the subsurface of the ground at a predetermined angle, in order to locate buried mines.

Prodding Probing the subsurface of the ground with a prodder or another tool, in order to locate buried mines.

Rate of processing The time required to process all the bags, divided by the number of bags processed.

Real-time Detection of explosives is carried out, with immediate response, during the passing-through the detector of the examined object (i.e. piece of luggage) or the walking-through a screening portal of a person.

Remote detection Detection of explosives from a certain distance, without having to bring the detector close to the examined object.

Retention time R_t The time required for a component to emerge from a GC (gas chromatograph) or HPLC (High performance liquid chromatograph) column.

Sampler A hand-held device which draws air through a removable cartridge, containing explosive-absorbing material.

Sampling To take a sample or specimen from a container, a piece of luggage, a parcel or a letter, in order to check for the presence of explosives.

Screening booth A walk-through booth for screening of persons for hidden explosives.

Screening portal Screening booth.

Screening rate The rate at which objects are screened for explosives.

Shaped charge A charge shaped so as to concentrate its explosive force in a particular direction.

Sheet Sheet explosive, usually Detasheet.

Signature Characterization of a certain element (e.g. nitrogen) by a radiation (e.g. γ-rays), having a certain energy.

Simulant A nonexplosive compound which has similar detection characteristics as an explosive, and can be used to test explosives detection systems.

Sniffer Detector based on the detection of vapor emanating from the explosive.

Surface-sampling detector A detector based on the detection of particles of explosives, left on the surface of the container holding the explosives.

Swabbing Removal of particles from a surface, with an adsorptive material.

Tagging Marking an explosive, during production, with an additive (tag, taggant) to make detection or identification easier. There are detection or identification tags.

Thermolabile Thermally unstable. Most explosives are thermolabile, which means that high temperatures may cause their decomposition.

Threat The possibility of the presence of an explosive in an examined object.

Tilt rod A post or pole normally attached to a fuse mechanism on top of a mine. Pressure against the tilt rod activates, by breaking or releasing mechanical retaining devices, the chain of fusing mechanism.

Throughput The number of pieces of luggage put through an explosive detector.

Unexploded ordnance, UXO Explosive ordnance, which has been primed, fused, armed, or otherwise prepared for use. It could have been fired, dropped, launched, or projected, yet remains unexploded, either through malfunction or design or for any other reason.

Vapor pressure The pressure exerted when a solid or liquid is in equilibrium with its own vapor. The vapor pressure is a function of temperature, increasing as the temperature increases.

Vapor transport The transfer of explosive vapor from the examined object to the detector.

Vapor generator A device for the production of explosive vapor, used for calibration of detectors.

Vapor trapping Trapping of explosive vapor, usually by adsorption, in order to concentrate it, before analysis by the detection system.

Volatility The ease by which a liquid or a solid changes into vapor.

Voxel The amount of a certain element (e.g. nitrogen) in a volume unit.

Walk-through The walking through a booth by a person for the detection of hidden explosives.

Z_{eff} Effective atomic number.

2

Vapor detection methods (sniffers)

2.1 INTRODUCTION

Vapor detection methods (in contrast to radiation-based methods) are non-invasive and measure traces of characteristic volatile compounds that evaporate from the explosive or are present on the container surface.

Most explosives have a low vapor pressure. Although the sensitivity of a sniffer depends largely on the vapor pressure of the compounds of interest, additional factors limit its sensitivity, such as efficiencies for vapor collection, transport of vapor, trapping, and collection [23]. As a result of these variables, the number of molecules available to the detector will be reduced further by a few orders of magnitude. The minimal detectable signal is generally evaluated using either a vapor generator or dilute solutions of pure explosives.

False positives are tested by tuning the detector to explosives of interest at concentrations which may be encountered in the environment, and observing the response, when the detector is exposed to a non-explosive compound [24].

False negatives are tested by turning the detector to a sample of explosive at the detection limit of the instrument, with a potential interferrant. If the detector fails to give a response, the experiment must be repeated using larger and larger amounts of explosives, until the detector gives a response. A new minimum detection level is then determined.

Typical interferrants for an explosives vapor detector are dirt, tobacco smoke, cigarette ash, perfumes, body odors, etc.

For an explosive vapor detector to be of practical use, it must, after having analyzed thousands of samples per day without ever coming in contact with an explosive, be able to detect an explosive at the detection limit level, without the explosive being adsorbed or absorbed on active interior surfaces of the instrument, and thus not reaching the detector.

An instrument may perform very well in a quick test, where every second, or third or even tenth sample is an explosive, but may fail totally in an airport environment, where only one in every hundred thousand samples may be an explosive.

2.2 TRAINED ANIMALS

Dogs and some other animals have a highly sensitive olfactory system. Dogs have been successfully used by law enforcement agencies to detect hidden illegal drugs. Similarly, dogs have been trained to sniff explosive vapors. The dog is the original vapor detector. For many purposes, it is still the vapor detector of choice. Dogs which have been used for explosives detection include German Shepherds, Labradors, Springer Spaniels and collies [25].

The olfactory epithelium, situated under the forehead of the dog, has been intensively studied. It has a large convoluted surface area: in the German Shepherd this area is about $100 \, cm^2$, compared to $3 \, cm^2$ for a human [25]. The surface of the olfactory epithelium is covered with olfactory cilia, which are attached to the receptors. These receptors are bipolar cells, the dendrites of which terminate in a small swelling from which project as many as ten or more cilia, which may be as long as $100 \, \mu m$ and $10^{-6} \, cm$ in diameter [26]. The cilia float in a mucus that bathes the olfactory epithelium, so that the receptors are in direct contact with the air when they lie on the surface of the film. At its proximal end, each primary receptor is connected by its own nerve fiber to the olfactory bulb, where the signals are processed before being sent to the brain. Odors dissolve in the mucus and trigger a transduction system within the cilia, which results in the generation of an electrical signal within the receptor, which in turn passes to the olfactory bulb. A good search dog can have a knowledge of at least 14 different types of odor, including drugs, human odors, and explosives. Moreover, since the dog's nares are completely separated by a septum, it has essentially a bilateral separation of olfactory stimuli, which allows it to detect the direction and/or location of the odor source [27].

The canine nose is an extremely sensitive molecular sniffer, able to detect vapors at concentrations three to five orders of magnitude lower than those discernible by people.

Dogs are trained on specific explosive ingredients instead of a specific product. A dog is trained to detect nitroglycerin instead of dynamite, so that the exact formulation is not important. When a dog is trained to detect a substance, it learns to discriminate between the vapor of that substance and other odors in the environment, by reacting to the compound or compounds that best help it to earn reinforcement from the handler [28]. With sufficient training, the compound or compounds whose detection most often results in reinforcement, become the 'odor detection signature' or 'olfactory detection signature'. The 'odor detection signature' of a nitroglycerin/nitrocellulose based smokeless powder was found to be determined largely by the three most abundant constituents of the powder: acetone, toluene, and limonene.

There are distinct advantages in the use of dogs for the detection of explosives: dogs are mobile. They can clear a large space, such as an auditorium or inspect a building floor-to-floor to assure the absence of explosives.

However, there are several problems associated with the use of detector dogs. Dogs frequently exhibit a decline in performance over time, after field work, which requires checking and constant retraining. It was found [27] that when performing a search task, dogs became tired after a time in the range 30–120 min, which suggested the need for two or more dogs at each location. In another case, a breakdown in working performance was observed in a male German Shepherd used in searches for explosives, after one year [29]. After a period of abstention from training followed by short sessions of training, the dog was again working as well as before.

The volume of flow at airports makes the use of dogs very expensive. Each dog requires an assigned handler for best performance, because the canine detector is really a specific dog–handler team. These factors result in high operational costs for the task of checking tens of millions of bags per year in national and international airports.

Like people, dogs have changing moods and show behavioral variations, which are difficult to monitor in a quantifiable way.

Small animals, such as rats and gerbils have also been used for the detection of explosives. The ability of a rat to detect different concentrations of TNT in air was investigated [30]. A rat was trained to press a bar when air containing TNT vapor was delivered, and to refrain from pressing that bar when air free of TNT was delivered. Adult male gerbils were trained to detect C-4, Deta Sheet and Semtex by means of an automated vapor injection system [31]. Each one of the gerbils was housed in a separate cage and had to actuate a lever when explosive vapor was delivered.

2.3 SAMPLING AND PRECONCENTRATION

A most important, but time-consuming, step during the use of explosives detectors is the sampling process. The traces of explosives must be collected and transported into the analytical system.

Adsorption effects must be considered when selecting materials used for explosive vapor transport. Explosives molecules were found to adhere to a variety of materials: Teflon, glass, Pyrex, quartz, nickel, stainless steel, gold, platinum, copper, fused silica, aluminum, and plastic tubing [32, 33]. At room temperature, nanogram quantities of these molecules were trapped before surface saturation. For glass and quartz tubes, heating in the range of 100–125°C was recommended to assure transport of explosives molecules. For nickel, stainless steel and Teflon, the recommended temperature was 150°C.

Vapor trapping devices, or preconcentrators, are used to trap and concentrate the explosives vapors before introducing them into the detection system. In such a way it is possible to increase the sensitivity of the detector by a factor which depends on the rate of concentration. This is especially important for explosives with relatively low vapor pressure. However, a vapor concentrator will increase the overall detection time.

Several types of vapor traps have been used [34]: volume traps, such as charcoal beds; surface traps, such as membrane filters; and solid surfaces, such as a metal surface which can be quickly heated. Flat plate geometry can be achieved by spiral winding a ribbon.

Volume traps are efficient, but difficult to desorb quickly.

Membrane filters are very efficient collectors of explosive vapors, but the pore size necessary for efficient collection results in a high pressure drop at the required flow rates. As a result, a large sampling pump is needed. Release of the collected vapor is done by heating the membrane.

It is difficult to release the collected vapor quickly, because it is difficult to heat a membrane quickly, a problem which exists also in the volume trap. As a result, the analysis time for this collection method is slowed down by the time required for heating.

A preconcentrator based on fullerenes as adsorbing material was designed [35]. Fullerenes are cage-like molecules of carbon in crystalline form: C_{60}, C_{70} and larger molecules. Fullerenes are soluble in benzene, toluene, and chlorinated aromatics and can readily be deposited as thin films on clean metal surfaces. Once deposited and thermally treated, the fullerenes adhere strongly to the metal surface and are difficult to remove. The coating is chemically and physically stable, up to temperatures of the order of $600°C$. Such thin films of fullerenes act as adsorbers and collectors of organic vapors. Because of their low thermal mass, the thin fullerene films can be heated rapidly by resistive electrical heating of their substrate and quickly desorb the vapors on their surface.

The concentrator consists of a glass tube containing a densely packed double helical coil of fine nichrome wire, which serves as substrate. The wire was coated by dipping it in a toluene solution of C_{60} and C_{70} fullerenes, and subsequent solvent evaporation by air drying at $70°C$. The collection efficiency for EGDN was about 40%. Similar coating of the coil with Tenax GC polymer, resulted in collection efficiency for EGDN of 100%.

While the fullerene concentrator tolerated repeated adsorption/desorption cycles in a rich oxygen environment, with little adverse effect on performance, the Tenax GC concentrator experienced thermal degradation and loss of trapping efficiency after repeated heating cycles under ambient air conditions [35].

Plastic explosives have an even lower vapor pressure than pure explosives. The vapor pressure of C-4 is lower than that of RDX, the difference being that C-4 contains polymeric binders and plasticizers, such as polyisobutylene, di(2-ethyl, hexyl)sebacate, and common motor oil. Suitable wrapping of the explosive will reduce even further the amount of vapor available to the detector, which might reach levels below the detection limit of the explosive detector. However, it may be difficult to clean all nearby surfaces, contaminated with explosives molecules, while the bomb was being prepared. Therefore, in addition to the conventional vapor sampling approach, we are looking now for traces of solid residue, microparticles of explosives, left unintentionally on various surfaces by a person who handled explosives. The sampler's function is now to remove particles of such explosives from surfaces, and gather them on a filter or a special collecting surface. The collector is then heated to vaporize the explosive particles, followed by transport of the vapor to the analytical system for detection [36]. Several approaches for the collection of particles have been examined [37].

- Mechanical agitation: a mechanical force is applied to the tested surface, followed by direct air sampling. It was found to be highly efficient for intact particles on all surfaces, but limited for oily residues on smooth surfaces.
- Swabbing: particles are removed with an adsorptive material. This time consuming method was found to be ideal for oily residues on smooth surfaces, but limited on rough surfaces.
- Vibration: sonic or mechanical vibration of the tested item was followed by air sampling. This method was successful for particles which are readily removable, but had a high dilution effect. There is a danger that strong forces may damage the tested item.
- Thermal desorption: heating of the tested item by infrared, laser, or other direct heating was used in order to vaporize the sample. This was followed by direct air analysis of the desorbed vapors. In addition to being a complex sampling system, there is a risk that thermolabile compounds, such as RDX, might decompose. There is also a risk to contents of the tested item.

A more common source of explosive particles are explosive residues deposited on surfaces in the form of hand and thumbprints. Direct touching of objects with the contaminated hands of an explosive handler is a most probable mechanism of transferring explosives traces to various surfaces. In such cases, a powerful vacuum cleaner, providing strong suction, was found to be an inefficient sampling method. Wiping or swabbing was much more effective [36].

2.4 GAS CHROMATOGRAPHY (GC)

In gas chromatography the sample (in the form of gas or volatile liquid) is introduced, through a heated injector, into a column, containing a stationary phase, usually a nonvolatile liquid held on a solid support. The sample is pushed through the column by a 'carrier gas' which constitutes the mobile phase. When a mixture, containing different components, is introduced into the column, these components will move through the column at different speeds, depending on the interaction of each component with the stationary phase. As a result, each component will emerge at a different 'retention time (R_t)', and thus a separation has been achieved. The emerging components are detected by a detecting device which produces a signal, the intensity of which is proportional to the concentration of the component in the mixture. Columns used today are mostly fused silica capillary columns of very small diameter, coated with a thin liquid film.

Two types of detectors have been used for the detection of explosives: the electron-capture detector (ECD) and the chemiluminescence detector. A third type, the surface acoustic wave (SAW) detector has also been suggested as an explosive detector.

2.4.1 Electron-capture detector (ECD)

A radioactive source of electrons (^{3}H or ^{63}Ni) ionizes the carrier gas (usually nitrogen), producing positive ions and thermal electrons. These electrons migrate to the anode and produce a standing current. When an electron-capturing compound (the sample) is introduced into the carrier gas, it captures some of these electrons to form negatively charged ions, thus reducing the standing current. The intensity of the reduced amount of current is proportional to the concentration of the sample. Electron-capture detectors have a fast response and are highly sensitive to most electron-capturing compounds, containing electronegative atoms, such as halogens or nitrogen. These compounds include nitroexplosives, but also conjugated carbonyls, alkyl halides and organometals. The ECD is virtually insensitive to hydrocarbons, alcohols, ketones, etc. Their specificity for explosives is low. The ECD is easily contaminated, but easy to clean. It is sensitive to water, therefore the carrier gas must be dry.

A GC–ECD explosives detector is shown schematically in Figure 2.1 [38]. It is a multicapillary chromatography system, which can handle both vapor and particle collected samples from cotton swabbing gloves.

2.4.2 Chemiluminescence detector

The chemiluminescence detector, also known as thermal energy analyzer (TEA) [39–42], is a nitrogen-specific detector. In the TEA detector, the sample is introduced into a pyrolyzer, where it decomposes to produce a nitrosyl

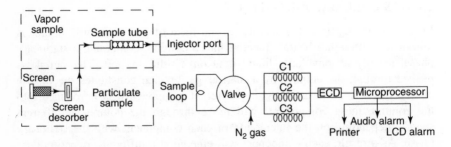

Figure 2.1 GC–ECD explosives detector. Reproduced from Nacson, S., et al., Second Explosives Detection Technology Symp., Atlantic City, NJ, 1996, p. 38, with permission.

radical NO•, either directly (in *N*-nitrosamines), or by catalytic reduction of the liberated NO_2 to NO (in nitroexplosives).

The three main groups of explosives, nitroaromatic compounds, nitramines, and nitrate esters, are pyrolyzed at a temperature of 350–800°C to NO• radicals according to Eqn 2.1:

$$=N-NO_2$$

$$-C-NO_2 \xrightarrow{\text{catalytic pyrolysis}} NO^\bullet + \text{product} \qquad (2.1)$$

$$-O-NO_2$$

The nitrogen oxide, together with other pyrolysis products, passes into a cold trap, held at a temperature of approximately −100°C. While the other pyrolysis products are retained in the cold trap, the NO• passes into a reaction chamber, maintained at about 3 torr, where it is oxidized by ozone according to Eqn 2.2, forming electronically excited nitrogen dioxide:

$$NO^\bullet + O_3 \longrightarrow NO_2{}^* + O_2 \qquad (2.2)$$

The $NO_2{}^*$ decays back to its ground state with emission of chemiluminescent light in the near-infrared region ($\lambda \cong 0.6$–$2.8\,\mu m$), according to Eqn 2.3:

$$NO_2{}^* \longrightarrow NO_2 + h\nu \qquad (2.3)$$

The intensity of the emitted light, detected by a photomultiplier, is proportional to the NO concentration and hence to the nitrocompound concentration. A red filter is placed in front of the photomultiplier to block any light with a spectral frequency higher than the near-infrared.

The TEA detector is very sensitive and selective. Its rejection ratio to hydrocarbons and *N*-containing organics is greater than $10^6 : 1$. The reasons for the high selectivity are as follows:

- Only compounds having NO_2 or NO functional groups will give a response.
- The reactive species must survive the −100°C cold trap.
- The reactive species must react with O_3 to produce chemiluminescent light in the narrow wavelength range of 0.6–2.8 µm.
- The reaction with O_3 must be rapid enough to occur while the effluent is in the reaction chamber, before being pumped away.

The TEA detector, although being much more specific for explosives than the ECD, is less sensitive by one to two orders of magnitude.

A GC–TEA explosive detector was developed with a two-column temperature-programmable high speed gas chromatograph, as shown schematically in Figure 2.2 [42]. The detector also includes an electrolytic gas generator to produce both hydrogen for the GC and oxygen for the ozone. The two columns SE_1 and SE_2 are operated in series with a 350°C pyrolyzer (PYRO-1)located between them. A second, 800°C pyrolyzer (PYRO-2), is located at

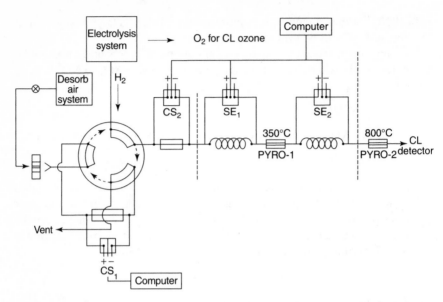

Figure 2.2 GC–TEA explosives detector: CL = chemiluminescence, PYRO = pyrolyzer, CS = cold spot, SE = chromatography column. Reproduced from Rounbehler, D.P., et al., First Int. Symp. on Explosive Detection Technology, Atlantic City, NJ, 1991, p. 703, with permission.

the end of SE_2. CS_2 is a temperature-programmed trapping cold spot that is used to inject the sample into the GC system. The hydrogen gas flow in both columns is about $40\,cm^3/min$. The computer controlled temperature program of the columns can be set at up to $100°C/s$.

The operation of the instrument is as follows: all the explosive samples pass through SE_1 and PYRO-1 before they enter SE_2. In PYRO-1, which is at 350°C, the nitrate ester and nitramine explosives will decompose to produce $NO^•$, which becomes now part of the carrier gas and passes through SE_2 without retention. The $NO^•$ gas from SE_1 will be detected by the chemiluminescence detector and displayed as the SE_1 chromatogram. Since nitroaromatic compounds do not decompose at 350°C, they will pass through PYRO-1 to SE_2, which, at this time is maintained at a low temperature; nitroaromatic explosives, if present, will condense and not pass through that column. At a preselected time, after SE_1 has completed its temperature program, SE_2 is temperature-programmed to separate the nitroaromatic explosives, which pass then through PYRO-2, which is at 800°C, where they are decomposed. The $NO^•$ produced from this pyrolytic decomposition is detected as a separate chromatogram (SE_2 chromatogram), following the SE_1 chromatogram. Figure 2.3 shows a typical chromatogram resulting from a vapor and particulates sample containing 5 pmol of each one of the explosives contained in the sample. Sampling was carried out by vacuuming vapors and

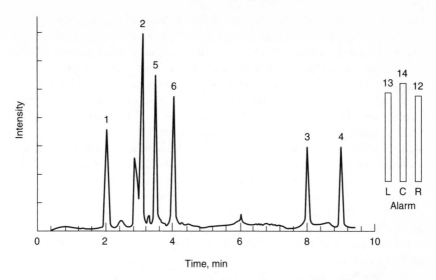

Figure 2.3 GC–TEA chromatogram from a vapor and particulates sample containing 5 pmol of each one of the following explosives: 1, NG; 2, EGDN; 3, DNT; 4, TNT; 5, PETN; 6, RDX. Reproduced from Rounbehler, D.P., et al., First Int. Symp. on Explosive Detection Technology, Atlantic City, NJ, 1991, p. 703, with permission.

particles from the hands and clothing of a person onto a metal foil sample probe at a rate of 2 L/s for a total of 20 s. A mixture of explosives (5 pmol each) with a flow of ambient air was additionally injected into the sample probe. The sample is then thermally desorbed into the analytical system by rapid heating of the collector coil, while simultaneously forcing air through it. The vapors from the sample can be subsequently recondensed onto a subambient cold spot CS_2 for later analysis. The time required for a complete analysis, including sample desorption, chromatographic separation and signal processing, is 18 s.

2.4.3 Surface acoustic wave (SAW) detector

2.4.3.1 Principle of operation [43]

An acoustic wave confined to the surface of a piezoelectric substrate material is generated and allowed to propagate. If matter is present on the same surface, then the wave and the matter will interact in such a way as to alter the properties of the wave (e.g. amplitude, phase, harmonic content, etc.). The measurement of changes in the surface wave characteristics is a sensitive indicator of the properties of the material present on the surface of the device. Applying a time varying electric field to the piezoelectric crystal will cause a synchronous mechanical deformation of the substrate with a coincident generation of an acoustic wave in the material. Proper selection of a single crystal orientation for the substrate will result in the acoustic wave propagation being constrained to the surface. The electric field is applied to the piezoelectric substrate by means of a thin metal foil transducer. A source of RF (radio frequency) excitation is then connected to the transducer and a surface wave is generated. The same type of transducer that is used to generate the surface wave can also be used to detect the surface wave. Thus, the construction of a simple surface acoustic wave delay line provides a means for the generation of a surface wave, a means for interaction with matter on the surface of the delay line, and a means for subsequent monitoring of changes in the wave resulting from the wave–matter interaction. Amplitude response has been found to be proportional to the pressure of the ambient gas and to the square root of the molecular weight of the ambient gas.

Quartz and lithium niobate SAW devices have been used for application in GC detectors [44]. Results showed that greater sensitivity could be obtained when using frequency rather than amplitude measurements.

2.4.3.2 SAW explosives detector

In the SAW based GC system, the SAW resonator crystal, operated at 500 MHz, is exposed to the exit of gas of a capillary column by a carefully positioned and temperature-controlled nozzle (Figure 2.4) [45]. When condensable analyte vapors impinge on the active area of the SAW crystal, a frequency shift occurs, proportional to the mass of the material condensing on the crystal surface. The frequency shift is dependent upon the mass and elastic constants of the material being deposited, the temperature of the crystal, and the chemical nature of the crystal surface.

The adsorption efficiency of each compound is a function of the crystal temperature, and by operating the crystal at different temperatures, the crystal can be made specific to a range of materials based upon their vapor pressure. A temperature of 0°C was chosen to ensure that most effluents, even volatile vapors, are trapped on the SAW crystal surface.

The SAW–GC detector operates as a true integrating device and as such provides information on the adsorption and desorption characteristics of each material which exits from the GC column.

The major elements of a GC–SAW vapor detection system are shown in Figure 2.5 [46]. The analysis is performed in two steps corresponding to the two positions of the GC rotary valve. In the sample position (shown), air to be tested passes through an optional inlet filter and through a loop trap, which contains an absorbent. Selection of sample time and flow rate determines the total amount of airborne vapors collected in the loop trap. The GC valve is rotated to its second position and the loop trap is rapidly heated by a capacitive discharge which causes trapped vapors to be transferred to the GC

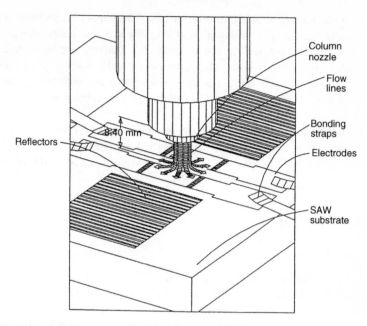

Figure 2.4 Surface acoustic wave (SAW) resonator sensor. Reproduced from Watson, G., et al., First Int. Symp. on Explosive Detection Technology, Atlantic City, NJ, 1991, p. 589, with permission.

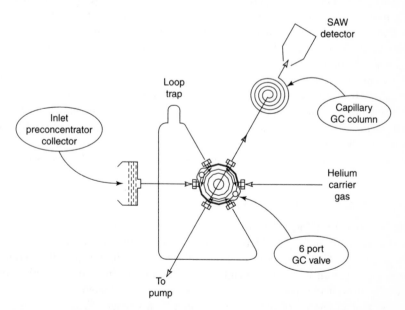

Figure 2.5 Schematic of GC–SAW system. Reproduced from Staples, E.J., et al., Pittsburgh Conf. on Analytical. Chemistry & Applied Spectroscopy, New Orleans, LA, 1998, with permission.

column, with the aid of a helium carrier gas. The vapors recondense on the inlet of a chromatographic column, held initially at a low temperature. After applying a heating gradient to the column, the analytes are separated, and as they elute from the column, they condense on the SAW crystal and are detected as frequency changes. The system was able to carry out a complete analysis in 10–15 s, when using a 1 m DB-5 capillary column (0.18 mm inner diameter).

Scale factors for frequency shifts for the SAW resonator system operating at 500 MHz, when exposed to TNT and RDX, were 9600 Hz/ng for TNT and 14 460 Hz/ng for RDX. Detection sensitivity was found to be in the picogram range.

2.5 MASS SPECTROMETRY

2.5.1 Introduction

Mass spectrometry is the field dealing with separation and analysis of substances according to the masses of the atoms and molecules of which the substance is composed.

The mass spectrometer consists of four main parts: the sample introduction system, the ionization source, the mass analyzer, and the detection and data acquisition system [47].

The mass analyzer and part of the detection system have to be under high vacuum. So does the ion source, in some of the ionization techniques.

The principle of mass analysis is that parameters of time and space of the path of a charged particle in a force field in vacuum are dependent on its mass-to-charge ratio (m/e).

The methods of separation and analysis can be divided into: methods based on geometric separation, and methods based on time separation.

The time separation method is based on the fact that ions having different m/e ratios have different times of flight and are thus collected one after the other. In the geometric separation method, ions having different m/e ratios are separated according to their geometric position at the collecting spot.

2.5.1.1 Magnetic sector mass spectrometer

The positive ions formed in the ion source are accelerated by a voltage, V, into a magnetic field, B, which is perpendicular to the direction of motion of the ion beam. Upon entering the flight tube between the poles of the magnet, each ion experiences a force at right angles to both its direction of motion and the direction of the magnetic field, and is therefore deflected.

The kinetic energy of the ion accelerated by a voltage, V, is given by:

$$\tfrac{1}{2}mv^2 = eV \tag{2.4}$$

The magnetic field exerts a centripetal force on the ion, resulting in its curving motion. Only ions of m/e ratio that have equal centrifugal and centripetal forces pass through the flight tube:

$$mv^2/r = Bev \qquad (2.5)$$

Where e = single charge of ion
$\quad\quad m$ = mass of ion
$\quad\quad r$ = radius of curvature of ion path
$\quad\quad V$ = accelerating voltage
$\quad\quad B$ = magnetic field strength
$\quad\quad v$ = velocity of ion

From Eqns 2.4 and 2.5 we obtain:

$$m/e = r^2B^2/2V \qquad (2.6)$$

Eqn 2.6 describes the principle of operation of the magnetic sector mass analyzer.

In most cases the ions are singly charged, therefore $e = 1$, so that for a given accelerating voltage, V, and a magnetic field, B, ions having different masses will follow trajectories having different radii: hence a mass separation is obtained. A mass spectrum is formed by scanning the magnetic field, B, (or the accelerating voltage, V), and having the separated ions pass through a slit and hit a detector (usually an electron multiplier connected to an amplifying and data system)

2.5.1.2 Time-of-flight (TOF) mass spectrometer

A time-of-flight mass spectrometer uses the differences in transit time through a drift region to separate ions of different masses. It operates in a pulsed mode, so ions are produced and extracted alternately in pulses. A voltage, V, accelerates all ions into a field-free drift region with a kinetic energy (see Eqn 2.4) $\frac{1}{2}mv^2 = eV$, where e is the ion charge and V is the applied voltage. The velocity of an ion with an m/e ratio is therefore:

$$v = (2eV/m)^{1/2} \qquad (2.7)$$

The transit time, t, through the drift tube is:

$$t = l/v = l(m/2eV)^{1/2} \qquad (2.8)$$

where l is the length of the drift tube.

2.5.1.3 Quadrupole mass spectrometer

The quadrupole analyzer consists of four parallel rod-shaped metal electrodes of hyperbolic cross section to which RF and d.c. voltages are applied

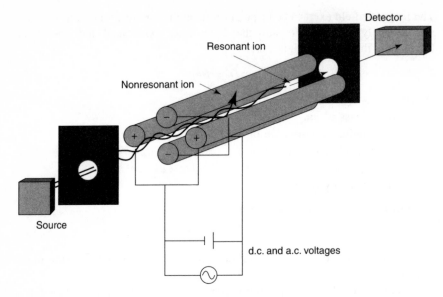

Figure 2.6 Quadrupole mass spectrometer.

(Figure 2.6). In practice four cylindrical rods are used. Two opposite rods are electrically connected and have an applied potential of $(U + V_0 \cos \omega t)$, and the other two rods, also electrically connected, have a potential of $-(U + V_0 \cos \omega t)$, where U is a d.c. voltage and $V_0 \cos \omega t$ is an RF voltage. The motion of an ion injected into the field in the z direction can be described by the Mathieu differential equations [48]:

$$m\, d^2x/dt^2 + 2e/r_0^2 (U + V_0 \cos \omega t)x = 0 \qquad (2.9)$$

$$m\, d^2y/dt^2 - 2e/r_0^2 (U + V_0 \cos \omega t)y = 0 \qquad (2.10)$$

$$m\, d^2z/dt^2 = 0 \qquad (2.11)$$

where r_0 is the radius of the four-rod system, m the mass of the ion, and e its charge.

The solutions of these equations designate oscillations performed by an ion in the x and y directions. Meanwhile, the ion proceeds with constant velocity in the z direction.

For a certain set of parameters, an ion of m/e ratio will continue an oscillating path through the quadrupole field until it hits the detector, while for ions of other mass-to-charge ratios, the oscillations will be unstable and the ions will hit the walls. The mass spectrometer is scanned by varying U and V_0, but keeping U/V_0 constant. As a result, ions with consecutive masses hit the detector, producing a mass spectrum.

2.5.1.4 *Ion trap (IT) mass spectrometer [49, 50]*

The ion trap consists of two end-cap electrodes that are normally held at ground potential with an interposed ring electrode to which a d.c. and RF voltage is applied so as to generate a quadrupole electrical field (Figure 2.7). The ring electrode is a single surface formed by an hyperboloid of rotation. The end-caps are complementary hyperboloids having the same conical asymptotes: z is an axis of cylindrical symmetry.

Ions are formed within the chamber by electron ionization from an electron beam admitted through a small aperture in one of the surfaces, or formed in an external ion source and injected into the ion trap.

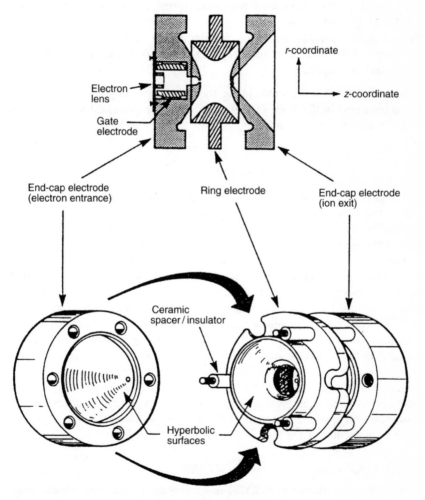

Figure 2.7 Schematic diagram of an ion trap. Reprinted with permission from Louris, J.N., et al., *Anal. Chem.*, **59**, 1677 (1987). © 1987 American Chemical Society.

The motion of an ion with a mass-to-charge ratio m/e can be described by the Mathieu differential equations:

$$m \, d^2 r / dt^2 + e/r_0^2 (U + V_0 \cos \omega t) r = 0 \qquad (2.12)$$

$$m \, d^2 z / dt^2 - e/z_0^2 (U + V_0 \cos \omega t) z = 0 \qquad (2.13)$$

where U is the d.c. voltage, $V_0 \cos \omega t$ the RF voltage, ω the angular frequency ($\omega = 2\pi f$), r_0 the radius of the cavity formed by the ring electrode (represents the trap size), z_0 the distance from the center of the trap to the end cap, m and e the mass and charge of the ion, respectively. For an ideal quadrupole field: $r_0^2 = 2z_0^2$. Solutions of Eqns 2.12 and 2.13 can be divided into two classes:

(a) r or z continues to increase with time, forming unbounded (and therefore unstable) trajectories.
(b) r and z are periodic with time, forming stable trajectories within the trap, provided the maximum distance of excursion from the center is less than r_0 and z_0, respectively.

Ions of given m/e are stable in the trap under operating conditions that can be summarized in the form of a stability diagram expressed in terms of the Mathieu coordinates a_z and q_z.

$$a_z = -8eU/mr_0^2 \omega^2 \qquad (2.14)$$

$$q_z = 4eV_0/mr_0^2 \omega^2 \qquad (2.15)$$

The region in the trap corresponding to simultaneous stability in both the r and z directions (even though it is plotted in terms of a_z and q_z only) can be shown in a stability diagram as shown in Figure 2.8. The shaded area represents solutions where the ion motion is bounded.

Mass analysis of the trapped ions is carried out by the mass-selective instability mode. In this mode, an RF voltage at a fixed chosen frequency is applied to the ring electrode. No d.c. voltage is applied ($U = 0$ and thus $a_z = 0$). This means that the locus of all possible values of m/e is on the q_z axis of the stability diagram. To record a mass spectrum, the RF voltage is increased with time, so that trapped ions are moved along the q_z axis until they become unstable at the boundary and exit the trap. Thus, as the RF voltage is raised, ions of increasing m/e values become successively unstable and are ejected in sequence through perforations in the end-cap and detected by an electron multiplier. The constant presence of about 1 mtorr of helium enhances the mass resolution, because the damping of the buffer gas keeps the ion trajectories close to the center of the trap.

2.5.1.5 Tandem mass spectrometry (MS/MS)

Tandem mass spectrometry (MS/MS)-in-space is a technique based on the combination of two mass spectrometers in tandem, with a collision cell between them (Figure 2.9) [51, 52]. The first mass analyzer separates the ions produced in the

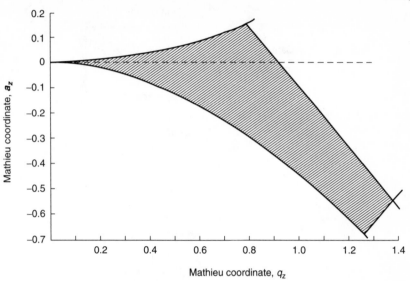

Figure 2.8 Stability diagram for ion trap mass spectrometer.

source. A precursor ion (or parent ion) is selected and focused into the collision cell. In the cell, the primary ion beam collides with an inert gas, such as helium, nitrogen, or argon, resulting in collision-induced dissociation (CID). The fragment ions (or daughter ions) thus produced in the collision cell are mass analyzed by the second mass analyzer and recorded. The CID mass spectrum (or daughter-ion mass spectrum) provides a 'fingerprint' of the primary ion.

An MS/MS instrument can perform in four modes of operation.

- Parent-ion mode. In parent-ion experiments the first mass analyzer is set to filter a single mass or a series of masses unique to the compound of interest. This mass selected parent ion undergoes collision-induced dissociation in the collision cell and is identified by its daughter-ion mass spectrum in the second mass analyzer. Parent-ion experiments permit identification of specific compounds in complex mixtures.

- Daughter-ion mode. In daughter experiments, the first mass analyzer is scanned, while the second one is set at a specific mass. This experiment identifies all parent ions that decompose to form a specific daughter ion.

- Neutral loss mode. In this type of experiment the two mass analyzers are set to detect a neutral loss. For example, the first mass analyzer can be scanned from m/e 117 to 317, at the same time as the second mass analyzer is scanned from 100 to 300. Since the two scanning functions are synchronized, the second mass analyzer will detect a daughter ion only if the focused parent ion loses the predetermined fixed mass. In this example, the neutral loss of 17 mass units may represent a series of nitroaromatic compounds losing OH.

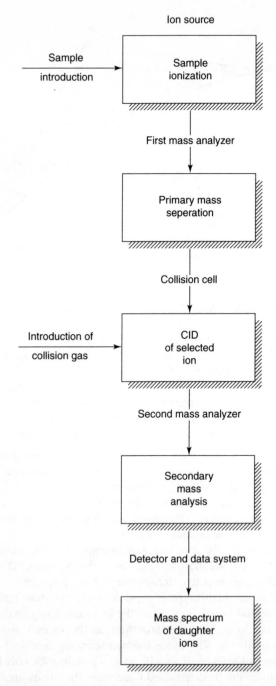

Figure 2.9 Schematic diagram of an MS/MS system (CID = collision-induced dissociation.)

- Single reaction monitoring (SRM). In this type of experiment, both first and second mass analyzers are set to specific masses. This mode of operation is practical when using MS/MS in combination with a chromatographic technique, such as GC/MS/MS or LC/MS/MS (LC = liquid chromatography). During the chromatographic run, it is possible to monitor a single transition between a specific parent ion and a specific daughter ion. This type of experiment can also be extended to multiple reaction monitoring (MRM)

There is a great variety of MS/MS-in-space combinations. The most popular one, which was also used in an explosive detector, is the triple-stage quadrupole mass spectrometer system (Figure 2.10) [53]. It consists of a tandem assembly of three quadrupoles. The first and third quadrupoles act as mass analyzers, while the second one, in which the collision cell is housed, acts as an ion focusing device, as only an RF voltage, without the d.c., is applied to it.

The ion trap mass spectrometer can be used as an MS/MS-in-time [49, 50]. In order to carry out MS/MS in a single analyzer instrument, three steps must be performed sequentially in parent-ion experiments: isolation of the parent ion, its dissociation into characteristic products, and their identification using a second stage of mass analysis:

- Isolation of the parent ion: after ionization, either externally, followed by ion injection into the ion trap, or internally, the parent ion is mass selected. This can be achieved by the mass-selective stability approach, that is, to apply combined d.c. and RF voltages on the end-caps to make all trapped ions, except the specified parent ion, unstable. Ions can also be isolated

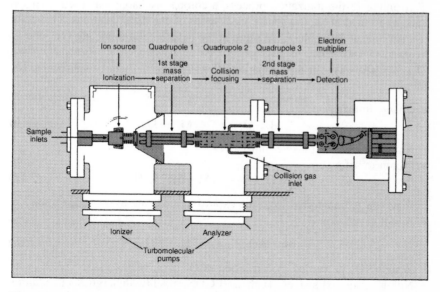

Figure 2.10 Triple quadrupole MS/MS system. Reproduced from Slayback, J.R.B. and Story, M.S., *Industr. Res. Dev.*, 129 (1981), with permission.

by resonantly ejecting all ions having lower and higher masses than the preselected parent ion.

- Daughter ions are formed from the parent ions by CID. The parent ions constantly undergo collisions with the helium in the ion trap. Normally these collisions involve relatively small energies, but if the translational energy of the parent ion is increased, the collisions may convert the translational kinetic energy to internal vibrational energy. If the parent ion acquires enough vibrational energy, one or more chemical bonds in the ion may be broken forming daughter ions. The translational kinetic energy of the parent ion can be increased using either one of two methods: nonresonant excitation or resonant excitation.
- The MS/MS-CID spectrum of the selected parent ion is then recorded by sequentially ejecting the daughter ions, using the mass-selective instability mode of operation.

Parent-ion and neutral loss experiments can also be carried out by ion trap MS/MS [54]. These modes of MS/MS have been applied in electrospray ionization-MS/MS of a series of explosives [55].

2.5.2 Detection of explosives by mass spectrometry

2.5.2.1 API-MS/MS detector

An explosives detector was built based on tandem mass spectrometry (MS/MS) with atmospheric pressure ionization (API) [56–58].

To initialize the ionization process, a corona discharge was used. Reagent ions are formed in the ion source via a series of chemical reactions. The major reagent ions at atmospheric pressure in the negative ion mode are O_2^- and its hydrates $O_2^- \cdot (H_2O)_n$.

The four pathways for the formation of explosives ions can be summarized as follows:

- Charge transfer

$$M + O_2^- \longrightarrow M^- + O_2 \qquad (2.16)$$

where M is the explosive molecule.
- Proton transfer

$$M + O_2^- \longrightarrow (M - H)^- + HO_2 \qquad (2.17)$$

A proton is transferred from the explosive molecule to the reagent ion.
- Addition reaction

$$M + NO_3^- \longrightarrow (M + NO_3)^- \qquad (2.18)$$

Where NO_3 is formed by a reaction of O^- with NO_2.
Nitrate ester explosives, such as EGDN, readily undergo such addition reactions.
- Dissociative charge transfer will produce fragment ions, such as NO_x^-

The variations in the composition of traces in the ambient air result in ambiguities in the composition of the reagent ions. Therefore, an additional reagent, methylene chloride, is introduced, forming a Cl^- ion. The chloride ion forms adduct ions with nitrate esters, such as PETN, NG and EGDN and with nitramines, such as RDX and HMX. In RDX, for example, addition of the chloride reagent produced a specific ion $(M + Cl)^-$ and increased the detection capability by over an order of magnitude. In the case of HMX, two different parent ion masses are used (m/z 331 and m/z 333) due to the different masses of the adduct ions, formed in the source with the isotopic $^{35}Cl^-$ and $^{37}Cl^-$ ions. For nitroaromatic explosives, such as TNT and DNT, charge transfer occurs from the O_2^- reagent ion, as shown in Eqn 2.16, since the target molecule has a higher electron affinity.

It was found that the presence of high concentrations of chloride ions suppresses the sensitivity of the detector for TNT and DNT, while if no chloride is present, the sensitivity of nitramine and nitrate ester explosives decreased by two orders of magnitude [59]. At approximately 0.1% concentration of chloride reagent gas, the intensities of all three groups of explosives were reduced only by 50% from the maximum response conditions. The concentration of methylene chloride in the mass spectrometer was therefore controlled and maintained at this level.

The triple quadrupole MS/MS system was used for mass analysis of ions formed in the ion source, fragmenting the selected parent ion in the collision cell and mass analyzing the produced daughter ions, which gave a specific 'fingerprint' of the explosive detected. Figure 2.11 shows the daughter-ion spectra of both the ^{35}Cl and ^{37}Cl adduct ions of HMX, at m/z 331 and m/z 333, respectively. Consecutive losses of 74 mass units are due to losses of CH_2NNO_2.

In practice, it is not necessary to acquire full spectra; once the fragmentation pathways of a compound have been identified, specific parent ion to daughter ion transitions (ion pairs) are selected, which are suitable for unique identification of the explosives of interest. In normal use, the tandem mass spectrometer is operated in the multiple reaction monitoring (MRM) mode. The selection of appropriate ion pairs for each species is important. The daughter ions must be present at high intensity levels to maximize sensitivity, but they must also be specific to maximize selectivity (that is to minimize false alarms). Ideally, the higher the number of daughter ions monitored for each compound of interest, the more selective the analysis. However, the rapid vaporization of the sample does impose constraints on the number of ion pairs which can be included in single sampling cycle. In general, two or three ion pairs are sufficient to produce optimum detection criteria.

Sampling was done by direct introduction of explosives vapor through air analysis, or by collecting involatiles or particles on a collector cartridge, which is then thermally desorbed to provide a vapor sample for analysis. A real-time-sampler could handle 12 items per minute. The sampling head contained

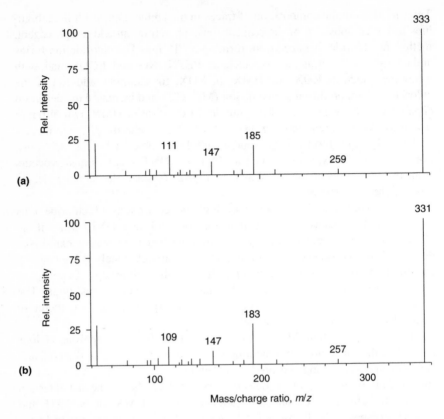

Figure 2.11 Daughter-ion mass spectra of: a, the $^{35}Cl^-$; and b, the $^{37}Cl^-$ adduct ions of HMX, at m/z 331 and m/z 333, respectively. Reproduced from Davidson, W.R., et al., First Int. Symp. on Explosive Detection Technology, Atlantic City, NJ, 1991, p. 653, with permission.

a rotating brush to maximize the removal of trace residues from the exterior surface of luggage and other items.

In airport environments, a rapid sampling system with immediate response is needed. Therefore, a real-time sampler (RTS) was developed [58–60], as shown in Figure 2.12. The operation of the sampler is as follows. Particles are removed from the surface to be examined, transported along a flexible tube, and deposited on a moving belt. The explosives are then thermally desorbed and the vapor is transferred into the ionization region for subsequent analysis. Ions formed in the ion source are mass analyzed in the MRM mode.

The collector belt is a stainless steel mesh about 3 cm wide, in the form of a continuous loop. The mesh, which has an opening size of less than 25 μm, is an air filter which removes suspended particles from a flow of air. The size of the particle removed depends on the opening size and surface properties

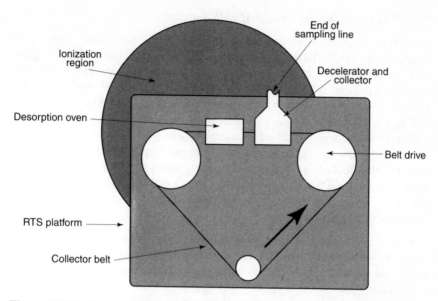

Figure 2.12 Real-time sampler. Reproduced from Sleeman, R. et al., Int. Symp. Contraband & Cargo Inspection Technology, Washington, DC, 1992, with permission.

of the mesh. The collector belt is driven continuously at a velocity of 5 cm/s. The sampler is normally used to analyze a flow of air containing suspended particles. The first step is to separate the sample particles from the carrier air stream. This separation takes place in the collection stage, which itself consists of three distinct components: an air inlet region, a particle collection zone and an air outlet region. Particle collection is a done in a chamber which is open to the atmosphere, via slots on three of its sides. The collector mesh (collector belt) is positioned in the middle of the collection zone. During operation, particle-laden air enters the inlet at high velocity, the air expands as it traverses the funnel-shaped inlet, and its linear velocity decreases accordingly. The air (containing the particles) strikes the filter mesh, and particles larger then the opening size of the mesh are retained by the collector. The air flows through the air outlet, which is connected to a vacuum pump. The next stage is desorption of the sample particles. The continuously-moving belt carries particles from the collection zone into the desorption region: this is a stainless steel oven maintained at a temperature in the range 230–300°C. As the mesh belt enters the oven, it is rapidly heated via both conduction and radiation: this evaporates or desorbs the compounds collected on the mesh. The sample vapors are carried by a flow of air into the ionization region of the mass spectrometer. Following ionization, the sample ions are mass analyzed and detected.

The clean-up stage consists of a moving mechanical brush (to dislodge large particles), followed by heating to 550°C and removal of residual ash from the mesh.

2.5.2.2 Glow discharge-ion trap MS/MS detector

Ionization in atmospheric sampling glow discharge ionization (ASGDI) is based on the establishment of a glow discharge in a region of reduced pressure, of about 1 torr, with ambient air as the discharge support gas [61–63]. A glow discharge is obtained by applying a voltage of 350–400 V between two electrodes. Electrons having an energy of several hundred eV are formed during the discharge. These electrons ionize molecules, thus forming slow electrons. Also, some of the fast electrons are slowed down by collisions with the ambient gas in the ion source. For species present at levels less than 1 p.p.m., the majority of negative ions are formed by electron capture. The main ions in ASGDI negative-ion mass spectra are M$^-$ and/or fragment ions. The main advantages of this type of ionization for explosives detection have been described [63] as follows:

- Simple and rugged design, without filaments or discharge needles, which can be operated continuously for months without maintenance.
- Air sampling rate of 1–5 mL/s with fast response and rapid pump-out of analyte.
- Less susceptible to matrix effects than API.

The ASGDI source was interfaced to an ion trap mass spectrometer for detection of explosives (Figure 2.13). MS/MS collision-induced dissociation of

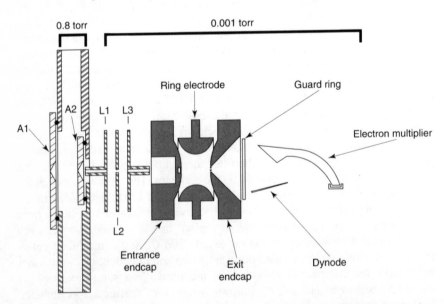

Figure 2.13 Schematic diagram of an atmospheric sampling glow discharge ionization (ASGDI) source interfaced to an ion trap mass spectrometer: A = aperture, L = lens. From McLuckey, S.A., et al., *Rap. Comm. Mass Spectr.*, **10**, 287 (1996). © 1996 John Wiley & Sons Ltd. Reproduced with permission.

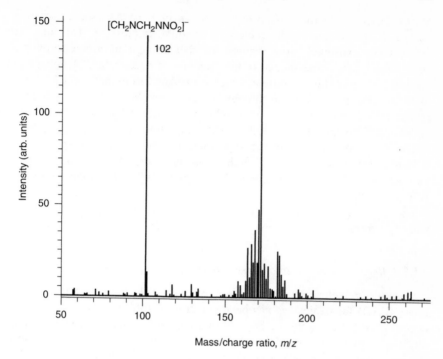

Figure 2.14 MS/MS spectrum of the $(M–NO_2)^-$ parent ion from a 500 fg sample of RDX. From McLuckey, S.A., et al., *Rap. Comm. Mass Spectr.*, **10**, 287 (1996). © 1996 John Wiley & Sons Ltd. Reproduced with permission.

characteristic explosives ions provided further identification of the explosives analyzed. High MS/MS efficiencies were obtained.

Figure 2.14 shows the MS/MS spectrum of the $(M–NO_2)^-$ parent ion from a 500 fg sample admitted into the heated region leading to the inlet of the ASGDI source. Most of the anion current produced in the ion source arises from major components in air. These species are usually much more abundant than the ions derived from the analyzed explosives, especially when the latter are present in the sub-p.p.b. level. The accumulation of these matrix ions can saturate the storage capacity of the ion trap. Therefore a mass-selective ion accumulation technique was used so that only ions in a preselected mass window were allowed to accumulate.

The detection limit of the explosives detector was found to be in the subpicogram range, with analysis time of 50 ms to 1 s.

2.5.2.3. API–time-of-flight mass spectrometer detector

A time-of-flight (TOF) mass spectrometer with atmospheric pressure ionization (API) has been suggested for detection of explosives [64, 65]. Air (containing the explosive vapor to be analyzed) is drawn into the API source where a corona discharge ionizes the molecules at ambient pressure. The ions are adiabatically expanded into a vacuum chamber through an orifice, forming a supersonic beam, perpendicular to the direction of acceleration. A voltage of 10–150 V is applied to the nozzle, which is insulated from the source housing, to focus the ions toward and through the skimmer. While the ion source is at atmospheric pressure, the nozzle and skimmer lead to pressures of 0.5 torr and 2×10^{-5} torr, in the first and second compartments, respectively.

The ions are pulsed from the supersonic ion beam into a perpendicular field-free drift tube, the time-of-flight mass analyzer, and are detected by a microchannel plate detector. Figure 2.15 shows a schematic diagram of the API–TOF MS detector.

The minimal detectable quantity of TNT was found to be 10 fg. A complete mass spectrum, with a resolving power of 500, was recorded in 200 µs.

2.5.2.4 Reversal electron attachment—mass spectrometer detector

A mass spectrometer with ion formation by reversal electron attachment detection (READ) for explosives has been constructed [66, 67]. The principle of operation is that electrons are brought to a momentary halt by reversing their direction with electrostatic fields. At this turning point the electrons have zero or near-zero energy. At electron energies below 10 meV, the cross section for electron attachment of certain groups of compounds, including explosives, is known to increase.

A schematic diagram of the READ apparatus is shown in Figure 2.16. It consists of an indirectly-heated cathode, F, from which electrons are extracted, accelerated by a five-element lens system, and focused into an electrostatic mirror. The mirror decelerates the electron beam to zero velocity at the reversal plane, R. The electron beam is square-wave modulated by fast switches S_1–S_3 with a nearly 50% duty cycle. Electron attachment to the explosives molecules takes place at R during one half of the pulse cycle. The resulting negative ions are extracted during the second half of the cycle (when the electron beam pulse is off), then focused, deflected by a 90° electrostatic analyzer, and further focused onto the entrance plane of a quadrupole mass spectrometer. The instrument was tested by having nitrogen at atmospheric pressure flowing over an ampule containing the explosives at room temperature. The nitrogen and explosives vapor pass through an adjustable molecular jet separator, which serves two functions: it reduces the pressure from the ambient 760 torr to 10^{-5} torr, which is required for the READ operation, and it enhances the concentration of the heavier molecular-weight explosives, relative to the N_2 carrier gas. The jet separator was heated to 100°C, which minimized adsorption of the explosives on the surfaces.

The READ mass spectrum of TNT contained a molecular ion and characteristic ions in the high-mass range. Sensitivity for TNT was found to be

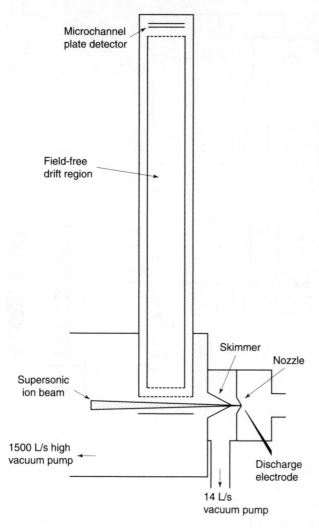

Figure 2.15 Schematic diagram of the API–TOF MS detector. Reproduced from Lee, H.G., et al., First Int. Symp. on Explosive Detection Technology, Atlantic City, NJ, 1991, p. 619, with permission.

better than 100 pg for an integration time of 10 s. The mass spectra of RDX and PETN contained mainly abundant low-mass ions: m/e 46 for RDX and m/e 46 and 62 for PETN.

2.5.2.5 Electron-capture negative ion mass spectrometry of explosives

Electron-capture negative ion mass spectrometry (ECNIMS) is known to be a very sensitive method for the detection and analysis of explosives [68, 69].

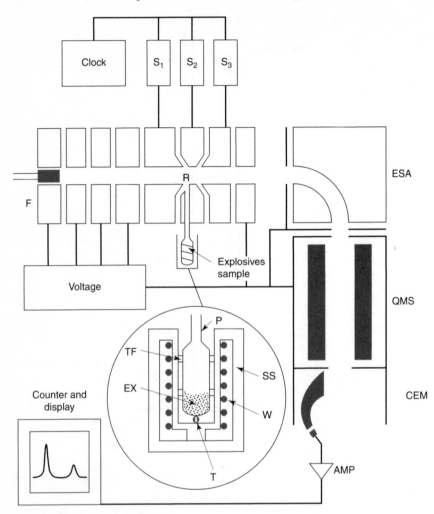

Figure 2.16 Schematic diagram of the reversal electron attachment detector (READ) apparatus. Reprinted by permission of Elsevier Science from Boumsellek, S., et al., *J. Am. Soc. Mass Spectrom.*, **3**, 243 (1993). © 1993 American Society for Mass Spectrometry.

Resonance electron capture can occur with the use of a buffer gas in the ion source, to moderate the electron energy or, at low pressure, by producing electrons at very low energies. This was achieved by building a trochoidal electron monochromator for low-energy electrons [70]. Figure 2.17 shows a schematic diagram of the electron monochromator–mass spectrometer system. Electrons emitted by a rhenium filament outside the monochromator are collimated and focused by four electrodes into a magnetic field produced by a pair of Helmholtz coils located outside the vacuum system. A small electric field

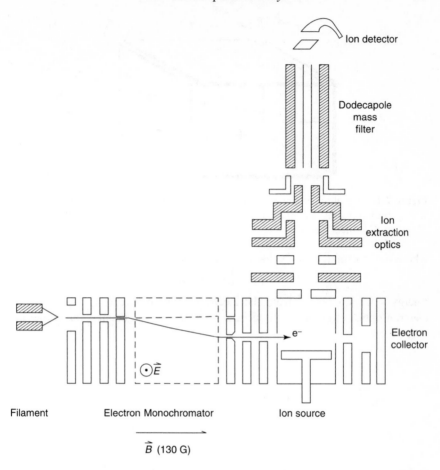

Figure 2.17 Schematic diagram of the electron monochromator/mass spectrometer system. Reprinted with permission from Laramee, J.A., et al., *Anal. Chem.*, **64**, 2316 (1992). © 1992 American Chemical Society.

of about 0.4 V/cm is established between a pair of vertical deflecting plates, oriented perpendicular to the electron beam. The applied potentials, referenced to the filament center, create a transverse electric field, E. This field, together with a magnetic field, B, parallel to the initial beam direction, results in the trochoidal motion of the electrons.

Electrons are deflected in a direction perpendicular to both the magnetic and electric fields and disperse according to their initial kinetic energies, v_0. The transverse displacement depends on the number of trochoids completed by the electrons during their residence time in the crossed field region (Figure 2.18), and is given by:

$$D = v_{\mathrm{d}}L(m/2v_0)^{1/2} \tag{2.19}$$

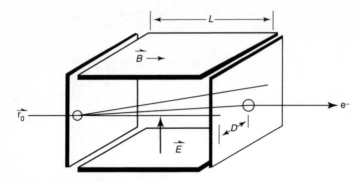

Figure 2.18 Operating principle of the trochoidal electron monochromator. Reprinted with permission from Laramee, J.A., et al., *Anal. Chem.*, **64**, 2316 (1992). © 1992 American Chemical Society.

where the trochoidal drift velocity is given by:

$$v_d = (E \times B)/B^2 \tag{2.20}$$

Energy selection results from the electron's time-of-flight. The assembly of electron optic components produces an intense electron beam, even at thermal energies.

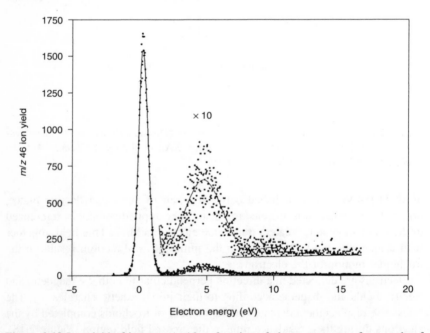

Figure 2.19 Ion yield of ion at m/z 46 as function of electron energy of a sample of Composition C-4. Reproduced from Laramee, J.A., et al., 5th Int. Symp. on Analysis and Detection of Explosives, Washington, DC, 1995, with permission.

An electron monochromator was also interfaced to a magnetic sector mass spectrometer [71]. Electron beams of 230 μA at kinetic energies of 0.03–60 eV were generated, producing ions with good peak shapes.

Differentiation between various explosives containing NO_2^- fragment ions, and between explosives and other nitrocompounds can be done by observing the electron energy at which this ion is formed [72]. The appearance potential (AP) for an ion, B^-, in the reaction:

$$AB + e^- \longrightarrow A^\bullet + B^- \qquad (2.21)$$

is given by the following equation:

$$AP(B^-) = BDE(A–B) - EA(B) + E_{Excess} \qquad (2.22)$$

where: BDE = bond dissociation energy
EA = electron affinity.

It is thus possible to differentiate between nitroaromatic explosives, such as TNT, containing a $C–NO_2$ bond, nitrate esters, such as PETN, containing a $O–NO_2$ bond and nitramines, such as RDX, containing a $N–NO_2$ bond.

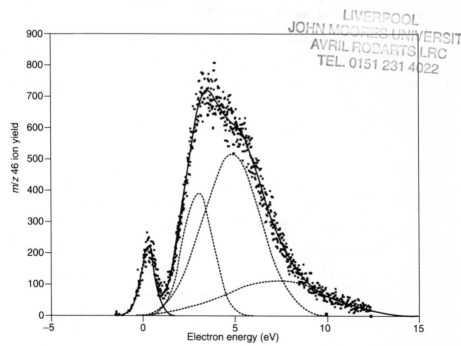

Figure 2.20 Ion yield of ion at m/z 46 as function of electron energy of a sample of TNT. Reproduced from Laramee, J.A., et al., 5th Int. Symp. on Analysis and Detection of Explosives, Washington, DC, 1995, with permission.

Composition C-4 (explosive ingredient is RDX) shows negative ion resonances at 0.27, 4.61, and 9.77 eV for the formation of the NO_2^- anion (Figure 2.19), while TNT shows a narrow resonance at 0.38 eV and a broad peak centered at 3.3 eV, which, when deconvoluted suggests substructure at 3.12, 4.92, and 7.35 eV (Figure 2.20). Not only do negative ion resonances possess different energies, but their peak shapes are also substantially different, which provides another criterion for differentiation of different explosives. Nonexplosive nitrocompounds can be differentiated from explosives. For example, the NO_2^- anion in 2-nitropropane, a component in tobacco smoke, has resonances at 0.68, 4.56, and 7.88 eV. The NO_2^- anion in nitrobenzene has resonances at 1.18, 3.52, and 5.04 eV.

A suggested explosive vapor detector would contain a sampler for vapor as well as particulates and the electron monochromator, interfaced to an ion trap mass analyzer. Preliminary tests of such a system showed that 20 pg of PETN particulate was successfully siphoned through a 0.010-inch capillary pipe to produce a response above background.

2.6 ION MOBILITY SPECTROMETRY (IMS)

2.6.1 Principles of operation [5, 73, 74]

The ion mobility spectrometer (IMS) consists of a sample inlet system, an atmospheric pressure ion source followed by an ion–molecule reactor, an ion-drift spectrometer, and a detector (Figure 2.21). Sample ions formed in the reactor are injected into the drift region by an applied electric field, where

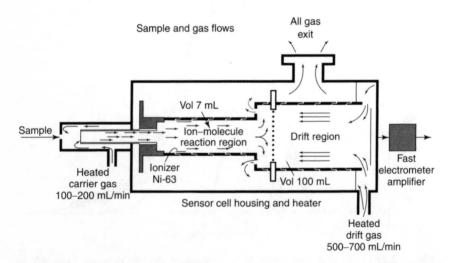

Figure 2.21 Schematic diagram of ion mobility spectrometer (IMS). Reproduced from PCP, Inc. (West Palm Beach, FL) brochure, with permission.

they are separated according to their mobility as they travel through a drift gas. The ion mobility spectrum consists of a plot of ion current as a function of drift time. The drift time depends on the ionic mass: heavier ions move at a slower speed and therefore have a longer drift time.

The drift velocity, v_d (cm/s), of an ion moving through an electric field, E(V/cm), is proportional to that field:

$$v_d = KE \tag{2.23}$$

where the proportionality constant, K, is the mobility of the ion in $cm^2\,V^{-1}\,s^{-1}$.

$$v_d = d/t \tag{2.24}$$

where d is the length of the drift region, and t the travel time. Hence:

$$K = d/(Et) \tag{2.25}$$

For a given temperature, T (in Kelvin), of the drift gas, and a pressure, P (torr), the mobility is given as reduced mobility, K_0, in the form of:

$$K_0 = K(273/T)(P/760) = (d/Et)(273/T)(P/760) \tag{2.26}$$

From Eqn 2.26 it can be seen that for a given set of electric field, temperature and pressure, the product of the reduced mobility and the drift time of an ion, $K_0 t$, is constant.

This means that the ratio of the reduced mobility of any two ions is independent of operating conditions and permits the use of an internal standard for calibration.

The most common electron source in IMS is a radioactive ^{63}Ni foil. Ions are formed in a carrier gas of nitrogen or air at atmospheric pressure, by 60 keV electrons emitted from the ^{63}Ni. The electrons lose energy during collisions with the carrier gas. The following sequence of reactions will occur:

$$N_2 + e \longrightarrow N_2{}^+ + 2e \tag{2.27}$$

$$N_2{}^+ + 2N_2 \longrightarrow N_4{}^+ + N_2 \tag{2.28}$$

Due to the presence of traces of water in the carrier gas and on the walls of the reaction chamber, charge transfer occurs from the $N_4{}^+$ ion to the water molecule as follows:

$$N_4{}^+ + H_2O \longrightarrow H_2O^+ + 2N_2 \tag{2.29}$$

Subsequently, H_2O^+ reacts with additional water molecules to produce ion clusters:

$$H_2O^+ + H_2O \longrightarrow H_3O^+ + OH \tag{2.30}$$

$$H^+(H_2O)_{n-1} + H_2O + N_2 \longrightarrow H^+(H_2O)_n + N_2 \tag{2.31}$$

The following reactions, leading to the hydrated nitric oxide ions, occur due to the presence of traces of oxygen:

$$N^+ + 2N_2 \longrightarrow N_3^+ + N_2 \tag{2.32}$$

$$N_3^+ + O_2 \longrightarrow NO^+ + O + N_2 \tag{2.33}$$

$$NO^+ + H_2O + N_2 \longrightarrow NO^+(H_2O) + N_2 \tag{2.34}$$

$$NO^+(H_2O)_{n-1} + H_2O + N_2 \longrightarrow NO^+(H_2O)_n + N_2 \tag{2.35}$$

The number of water clusters and hydrated nitric oxide ions, n, is a function of the temperature and the partial pressure of water in the gas. For example, at 25°C, 700 torr, and 5 torr partial water pressure (21% relative humidity), the clusters contain 5–8 water molecules.

H_2O^+ ions will react with NH_3 contaminants from the atmosphere, present in the reactor as follows:

$$H_2O^+ + NH_3 \longrightarrow NH_4^+ + OH \tag{2.36}$$

$$NH_4^+ + H_2O + N_2 \longrightarrow (H_2O)NH_4^+ + N_2 \tag{2.37}$$

Three types of ion–molecule reactions have been found to form positive product ions in IMS:

- Proton transfer reactions can occur if the proton affinity of the sample molecule is greater than that of the reagent ion:

$$(H_2O)_n H^+ + M \longrightarrow MH^+ + nH_2O \tag{2.38}$$

When introducing NH_3 into the reactor, an abundant reagent ion of the type $(H_2O)_n NH_4^+$ will be formed, which will transfer a proton to molecules, such as nitrogen-containing compounds, having a greater proton affinity than ammonia:

$$(H_2O)NH_4^+ + M \longrightarrow MH^+ + NH_3 + H_2O \tag{2.39}$$

- Charge transfer reactions have been observed between the $(H_2O)_n NO^+$ reagent ion and benzene derivatives:

$$(H_2O)_n NO^+ + M \longrightarrow M^+ + NO + nH_2O \tag{2.40}$$

- Electrophilic addition reactions of NO^+ to nitroaromatic compounds:

$$(H_2O)_n NO^+ + M \longrightarrow MNO^+ + nH_2O \tag{2.41}$$

In the negative-ion mode, when using nitrogen as both drift and carrier gas, the reagent species are the thermal electrons. When using air, the reagent species is O_2^-. Additional reagent ions are $(H_2O)_n O_2^-$, when traces of water are present in the air. The major reactions leading to the formation of negative ions are:

Electron capture:

$$AB + e^- \longrightarrow AB^- \qquad (2.42)$$

Dissociative electron capture:

$$AB + e^- \longrightarrow A^- + B \qquad (2.43)$$

The addition of a reagent gas will produce, through ionization by electrons, a reagent ion, R^-, which will undergo ion–molecule reactions with the sample molecule M:
Electron attachment:

$$M + R^- \longrightarrow M^- + R \qquad (2.44)$$

Proton abstraction:

$$M + R^- \longrightarrow (M - H)^- + HR \qquad (2.45)$$

Attachment of reagent ion:

$$M + R^- \longrightarrow (M + R)^- \qquad (2.46)$$

2.6.2 IMS detection of explosives

For the detection of explosives, three gases are introduced in the reaction chamber: dry air at atmospheric pressure, as the carrier gas, hexachloroethane (C_2Cl_6), as the reagent gas, and 4-nitrobenzonitrile, as the calibration gas. The reagent ion Cl^- will undergo ion–molecule reactions with an explosive molecule as outlined in Eqns 2.44, 2.45 and 2.46.

More than one species will be generated for some of the explosives, as can be seen in Table 2.1 [75]. These results were obtained when using the

Table 2.1 Characteristic ions in IMS for some common explosives. From Fetterolf, D.D., et al., *J. Forensic Sci.*, **38**, 28, © 1993 ASTM. Reprinted with permission

Peak	Proposed species (negative ions)	Mass	Drift time (ms)	Reduced mobility, K_0 ($cm^2\,V^{-1}\,s^{-1}$)	Limit of detection, LOD
TNT	TNT − H	227	14.52	1.451	200 pg
RDX-1	RDX + Cl	257	15.19	1.387	200 pg
RDX-2	RDX + NO$_3$	284	16.03	1.314	800 pg
RDX-3	RDX + (RDX + Cl)	479	22.22	0.948	1 ng
PETN-1	PETN − H	316	17.37	1.213	80 ng
PETN-2	PETN + Cl	351	18.40	1.145	200 pg
PETN-3	PETN + NO$_3$	378	19.08	1.104	1 ng
NG-1	NG + Cl	262	15.73	1.339	50 pg
NG-2	NG + NO$_3$	289	16.50	1.275	200 pg
NO$_3$	H$_2$O + NO$_3$	100	10.93	1.927	200 pg

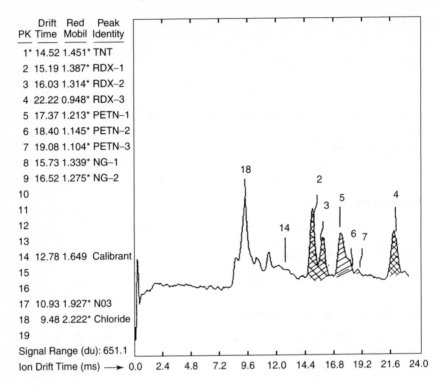

PK	Drift Time	Red Mobil	Peak Identity
1*	14.52	1.451*	TNT
2	15.19	1.387*	RDX–1
3	16.03	1.314*	RDX–2
4	22.22	0.948*	RDX–3
5	17.37	1.213*	PETN–1
6	18.40	1.145*	PETN–2
7	19.08	1.104*	PETN–3
8	15.73	1.339*	NG–1
9	16.52	1.275*	NG–2
10			
11			
12			
13			
14	12.78	1.649	Calibrant
15			
16			
17	10.93	1.927*	N03
18	9.48	2.222*	Chloride
19			

Signal Range (du): 651.1

Ion Drift Time (ms) ⟶ 0.0 2.4 4.8 7.2 9.6 12.0 14.4 16.8 19.2 21.6 24.0

Figure 2.22 Ion mobility spectrum of a postblast sample containing PETN and RDX. From Fetterolf, D.D., et al., *J. Forensic Sci.*, **38**, 28 (1993). © 1993 ASTM: reprinted with permission.

following operating conditions: drift and inlet temperatures were 95°C and 215°C, respectively; drift, sample, and exhaust flow rates were 350, 300, and 650 mL/min, respectively. Figure 2.22 shows the ion mobility spectrum of a sample containing RDX and PETN residues [75].

The relative intensities of the multiple ion peaks are influenced by concentration effects and by the stability and composition of commercial explosives [76]. The $(M + Cl)^-$ and $(M + NO_3)^-$ ion peaks are separated by less than 1 ms. In pure explosives $(M + Cl)^-$ is usually the stronger peak, and $(M + NO_3)^-$ may only appear as poorly resolved shoulder. In commercial explosives, the intensity of the $(M + NO_3)^-$ peak often increases and may be comparable to that of the $(M + Cl)^-$ peak. Figure 2.23 shows the ion mobility spectra of a series of commercial explosives.

In explosives that contain ammonium nitrate, chloride ions are suppressed and $(M + NO_3)^-$ becomes the strongest or only peak of the explosive molecule, M. That occurs because of the increasing amount of nitrate ions in the reaction

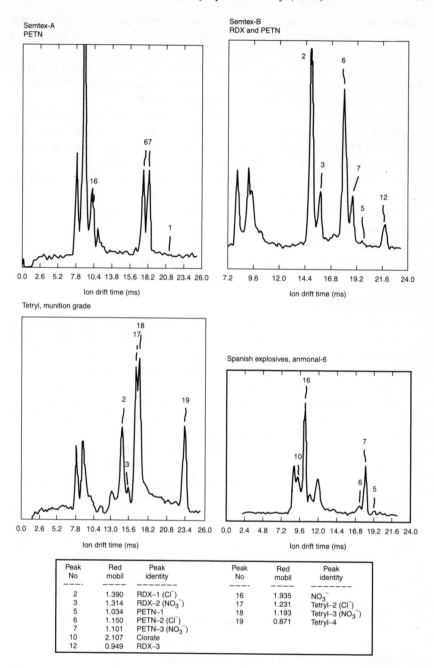

Peak No	Red mobil	Peak identity	Peak No	Red mobil	Peak identity
2	1.390	RDX–1 (Cl⁻)	16	1.935	NO₃⁻
3	1.314	RDX–2 (NO₃⁻)	17	1.231	Tetryl–2 (Cl⁻)
5	1.034	PETN–1	18	1.193	Tetryl–3 (NO₃⁻)
6	1.150	PETN–2 (Cl⁻)	19	0.871	Tetryl–4
7	1.101	PETN–3 (NO₃⁻)			
10	2.107	Clorate			
12	0.949	RDX–3			

Figure 2.23 Ion mobility spectra of a series of commercial explosives. Reproduced from Danylewych-May, L.L., et al., *Advances in Analysis and Detection of Explosives*, Dordrecht, 1993, p. 385, with kind permission from Kluwer Academic Publishers.

chamber. Nitrate ions compete effectively with chloride ions in the IMS ionization process, because the nitrate electron affinity (3.9 eV) is higher than that of chloride (3.6 eV).

A flow-through particle collection approach was found to be the most effective and simplest method for continuous in-line sample collection, in combination with a reel-to-reel tape drive system [76].

A hand-portable gas chromatography/ion mobility spectrometry (GC/IMS) instrument has been developed for detection of toxic chemicals, including explosives [77]. The instrument includes an automated vapor sampling system, a high speed capillary GC (with a 5 m column) and an IMS.

An IMS detector has been designed to detect traces of explosives on documents. This concept is based on the assumption that people who have handled explosives, might transfer explosive residues from their finger tips onto documents, such as a passport or boarding card.

2.7 INFRARED (IR) SPECTROSCOPY METHODS

2.7.1 Introduction

The infrared part of the electromagnetic spectrum extends from 10 000 to 5 cm^{-1}. This broad range is further subdivided into the near-IR (10 000–5000 cm^{-1}), mid-IR (5000–200 cm^{-1}) and the far-IR (200–5 cm^{-1}). The interaction of this radiation with matter excites vibrational transitions. When IR absorption occurs during a fundamental type of transition, a radiation frequency matches one of the vibrational frequencies of the absorbing molecules. In addition to the frequency match there is another requirement that must be met, which involves the manner in which the molecule is able to absorb the radiation energy: in order for any IR absorption to occur the molecular vibration must cause a change in the molecular dipole moment. The dipole moment is defined in the case of a simple dipole, as the magnitude of either charge of the dipole, multiplied by the charge spacing. In the near-IR, harmonics and combination transitions are observed, while in the far-IR, rotational transitions are observed in the gas phase. In the mid-IR, region referred to as the fingerprint region, transitions associated with fundamental molecular vibrations are observed. This part of the spectrum can be used for detection and analysis of explosives, because absorptions due to the NO_2 group can serve as a characteristic signature [78]. The absorptions corresponding to the NO_2 group vibrations, besides having well-defined group frequencies, are the strongest in the IR spectra of explosives.

2.7.2 Infrared absorption of explosive vapors

Infrared absorption spectra and band strengths were obtained for TNT, RDX and PETN [79]. The compounds were introduced into a small sample arm on the side of the main cell and both cell and sample container were heated separately. The side arm was held at a temperature of 20°C below that of the main cell body, to ensure that an equilibrium vapor pressure was maintained. The cell was maintained at a pressure of 10^{-4} torr. Spectra were measured over the temperature range 340–420 K. The spectrometer used was a Bio-Rad FTS60A, using the internal ceramic mid-IR source. The detector was a liquid nitrogen-cooled HgCdTe detector. Figures 2.24, 2.25 and 2.26 show the IR absorption spectra of TNT, RDX, and PETN, respectively. TNT vaporizes and condenses reversibly, and the spectrum contains the symmetric and antisymmetric NO_2 stretches at 1349 and 1559 cm^{-1}, respectively. In RDX and PETN, decomposition products are evident in the absorption spectrum, especially at higher temperatures. The NO_2 stretching bands can be readily identified, along with several additional features, but absorption from the decomposition products dominates at the higher temperatures.

Band strengths were found to be proportional to vapor pressure. A 5–10% less steep temperature dependence of the band intensities, as compared to the vapor pressure, was due to some band broadening at higher temperatures.

2.7.3 Infrared detection of explosives

The detector is composed of three main parts: sampling, catalytic decomposition of the explosives into a characteristic set of stable molecular fragments

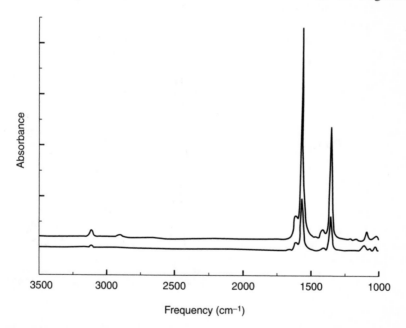

Figure 2.24 Infrared absorption spectra of TNT vapor at 120°C (lower trace) and 140°C (upper trace). Reprinted from Janni, J., et al., *Spectrochim. Acta A*, **53**, 1375, © 1997, with permission from Elsevier Science.

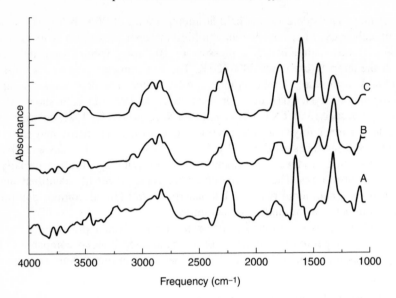

Figure 2.25 Infrared absorption spectra of RDX vapor at: A, 150°; B, 170°, and C, 200°C. Reprinted from Janni, J., et al., *Spectrochim. Acta A*, **53**, 1375, © 1997, with permission from Elsevier Science.

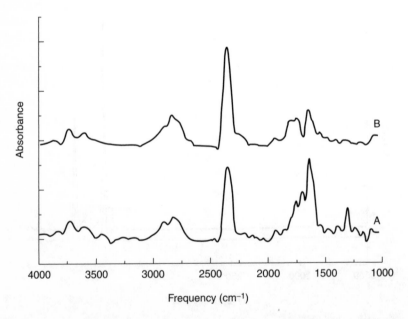

Figure 2.26 Infrared absorption spectra of: A, PETN vapor at 140°C, and B, PETN plus decomposition products at 160°C. Reprinted from Janni, J., et al., *Spectrochim. Acta A*, **53**, 1375, © 1997, with permission from Elsevier Science.

and desorption of the fragments for analysis by a tunable diode laser spectrometer [80].

The explosive molecules are adsorbed on the inner walls of a small diameter cell, after having been flushed from a preconcentrator (or from a vapor generator), into the cell in a stream of helium. The explosive molecule has to be adsorbed strongly enough so that decomposition is faster than desorption under the temperature, pressure, and gas-flow conditions that are used. Because TNT, RDX and PETN are low-vapor-pressure polar molecules, this requirement is generally easy to meet.

A second catalytic decomposition requirement is for the explosive to decompose into the selected fragments in acceptable yields for detection. These fragments are NO, NO_2, and N_2O. An ideal catalyst should generate large yields of these fragments and should also provide capability for distinction between nitramines, nitrate esters and nitroaromatic compounds.

The catalyst cell is a 12 cm stainless steel tube having a 0.294 cm inside diameter. Several coatings were used as catalysts: CeO_2, NaUSY zeolite, $LiClO_4$, and Al_2O_3–CeO_2. It is necessary to heat transfer lines to avoid adsorption, but without causing decomposition of the explosives. It was found that temperatures of transfer lines should be 170–200°C for RDX, and 140–160°C, for PETN.

The detection of the characteristic decomposition products is carried out by frequency modulation spectroscopy (FMS) with lead-salt diode lasers. A diode laser is modulated at radio frequencies (RF) by modulating its injection current. The modulated laser beam is directed through a sample cell containing the absorbing molecules. The resulting absorption converts some of the laser frequency modulation into amplitude modulation. This amplitude modulation can easily be detected by using a photodiode of suitable bandwith and standard RF signal-processing techniques. The advantage in using RF modulation and detection is that the signal of interest occurs in a frequency range where the laser has very low noise.

Detection limits in the p.p.b. to p.p.t. range may be obtained by using midinfrared lead-salt diode lasers that emit in the fundamental molecular absorption bands. These lasers operate only at temperatures below 100 K, and require cryogenic cooling.

The most intense N_2O, NO, and NO_2 lines are at 4.57 μm, 5.25 μm, and 6.24 μm, respectively, of the spectrum. The FMS detection system is shown in Figure 2.27. It consists of three main subsystems.

- Three diode lasers, their controllers, and three infrared detectors.
- The RF circuitry that modulates the lasers and detects the FMS signals.
- The data-acquisition system that controls the valves and heaters and acquires and processes the signals.

The three modulation frequencies for the N_2O, NO, and NO_2 lasers are 200 ± 2.5, 200 ± 5.5 and 200 ± 7.0 MHz, respectively.

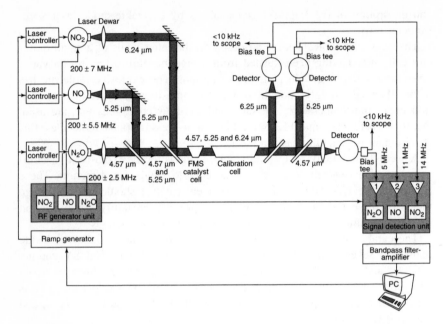

Figure 2.27 Frequency modulation spectroscopy (FMS) detection system. Reproduced with permission from Riris, H., et al., *Appl. Optics*, **35**, 4694, © 1996 Optical Society of America.

A decomposition yield of 0.8 (0.8 N_2O molecules for each RDX molecule) led to a lower detection limit of 5–10 pg for RDX. In a field instrument, the decomposition yield will, of course, be dependent on the explosive packaging and the actual vapor that can be sampled.

2.8 BIOLUMINESCENCE

2.8.1 Basic principles [81]

A light accompanying chemical reaction that occurs in a living system, or is derived from one, is known as bioluminescence (BL). BL methods are very sensitive because it is easy to measure low levels of light emission. In practice, sensitivity is usually limited by reagent purity, rather than by light-measuring capability.

A widely studied bioluminescence system is the firefly system, summarized in the following scheme:

$$LH_2 + E + ATP + Mg^{+2} \longrightarrow E{\bullet}LH_2{\bullet}AMP + MgPPi \qquad (2.47)$$

$$E{\bullet}LH_2{\bullet}AMP + O_2(g) \longrightarrow \{Oxyluciferin\}^* + AMP + CO_2 + H_2O \qquad (2.48)$$

$$\{Oxyluciferin\}^* \longrightarrow Oxyluciferin + h\nu \quad \lambda_{max} = 562\,nm \qquad (2.49)$$

where LH_2 = luciferin (substrate)

   E = luciferase (enzyme)

ATP = adenosine triphosphate

AMP = adenosine monophosphate

PPi = inorganic pyrophosphate

The minimum detection limit of this reaction is 0.1–1.0 pmol, with linearity of response extending five orders of magnitude. The BL efficiency (i.e. number of photons emitted per number of molecules reacting) of this reaction is one.

Another bioluminescence system is the bacterial system, summarized by the following reactions:

$$FMNH_2 + O_2 + E + RCHO \longrightarrow E\cdot FMN^* + RCOOH + H_2O \quad (2.50)$$

$$E\cdot FMN^* \longrightarrow E\cdot FMN + h\nu \qquad \lambda = 492\,nm \qquad\qquad (2.51)$$

where FMN = flavine mononucleotide.

RCHO = long chain aldehyde.

The BL efficiency of this reaction is about 0.05. The intensity of bacterial bioluminescence in this reaction is a function of the chainlength of the aldehyde. Bacterial BL is highly specific for FMN. The relationship between light output and $FMNH_2$ concentration is linear from 1.0×10^{-4} to 1.0 μg/mL.

The two most popular sources of the bacterial luciferase (E) are the bacteria *Photobacterium fischerii* and *Achromobacter fischerii*. The luciferin, FMN, may be obtained from virtually any living system. The luciferase is generally used at a concentration of 1.0 μg/mL in 0.05 M THAM (tromethamine) buffer (pH 7.4). Dodecylaldehyde complexed with bisulfite is widely used as the required aldehyde.

2.8.2 Bioluminescence detection of explosives

A detection system for TNT vapor, utilizing an enzyme that catalyzed the reduction of TNT molecules, has been developed [82]. Figure 2.28 shows a bioluminescent detector module.

A bioluminescent signal is obtained from a series of chemical reactions occurring in the detector module. Two primary reactions occur: one reaction with TNT absent and the second with TNT present.

The reagents react in the TNT_{ase} reaction cell as follows:

$$NADH + FMN + H^+ \longrightarrow NAD^+ + FMNH_2 \qquad (2.52)$$

where NADH is the reduced form of nicotinamide adenine dinucleotide and FMN is flavine mononucleotide. The reduction product $FMNH_2$ undergoes a bioluminescent reaction as follows:

$$FMNH_2 + RCHO + O_2 \longrightarrow FMN + RCOOH + H_2O + h\nu \qquad (2.53)$$

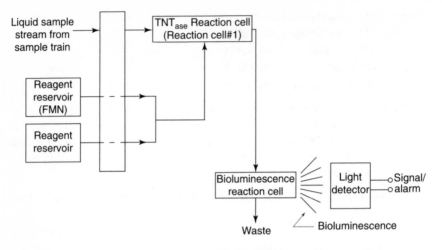

Figure 2.28 Bioluminescent detector module. Reproduced from Boncyk, E.M., 3rd Int. Symp. on Analysis and Detection of Explosives, Mannheim-Neuostheim, 1989, p. 40-1, with permission.

where RCHO is a long chain aldehyde. This reaction is catalyzed by an oxidoreductase and by bacterial luciferase. Light intensity is measured by a photomultiplier. In this system the reaction is continuous, because $FMNH_2$ is continuously replenished.

The second reaction, occurring when TNT molecules are present, is the reduction of TNT, promoted by the catalytic action of the TNT enzyme (TNT reductase) as follows:

$$M_{(TNT)} + NADH + H^+ \longrightarrow MH_{2(TNT)} + NAD^+ \qquad (2.54)$$

This reaction also consumes NADH molecules, and therefore, competes with the reaction described by Eqn 2.52, leading to a proportional decrease in the intensity of the bioluminescent reaction, which was found to be inversely proportional to the TNT concentration in the liquid stream.

The sample train performs the gas-phase sampling, preconcentration, and detector module interface functions. The system operates in an automated batch mode over a specific time period. A sample air stream passes continuously through the sample train. Simultaneously, a recirculating liquid water stream flowing in the sample train, extracts a large fraction from the TNT molecules from the air stream and stores and accumulates them in liquid solution. At the end of the sampling period, the liquid solution is removed from the sample train and is transported to the detector module for analysis. In effect, a gas-phase sample is transformed into a liquid sample to interface with the liquid-phase detector module, and is at the same time preconcentrated.

The fabrication of the TNT$_{ase}$ and bioluminescence reaction cells is accomplished with a heterogeneous reactor bed onto which the TNT$_{ase}$ enzyme is efficiently immobilized in the TNT_{ase} reaction cell, while the bacterial luciferase and oxidoreductase enzymes are co-immobilized in the bioluminescence reaction cell.

The system could detect a minimum concentration of 0.25 p.p.t. ($\times 10^{-12}$ g/cm^3) of TNT in air in a total analysis time of 20 min, including preconcentration.

2.9 LASER OPTOACOUSTIC SPECTROSCOPY

2.9.1 Physical principles of operation

The optoacoustic effect was discovered by Alexander Graham Bell, over 100 years ago [83]. He noticed that focused intensity modulated light falling on a solid substance could produce an audible sound. He later noticed the same effect in gases and liquids.

Photoacoustic spectroscopy requires a source of pulsed radiation, a photodetector to measure the power level, a cell to contain the gas of interest, a transducer to convert pressure waves into electrical signals, and a measuring device to monitor the ratio of the pressure transducer output and the radiation intensity.

Photoacoustic detection is performed in the following way [83–86]. The excitation source, usually an IR laser, is tuned to a wavelength, which is known to be absorbed by the molecules of interest. The radiation is focused into a cell containing a small partial pressure of the gas of interest in a buffer gas (usually air) at atmospheric pressure. A photodetector monitors the level of input radiation. A small part of the input energy is absorbed by the gas of interest and stored as vibrational energy. Collisions with the buffer gas converts the absorbed energy into translational kinetic energy of the entire gas mixture, resulting in increase in temperature and pressure inside the cell. Modulation of the incident radiation produces a periodic temperature rise, resulting in the generation of a periodic pressure wave which can be detected as sound wave by a sensitive microphone, located within the cell.

The system thus performs a measurement of incident and transmitted intensity, from which absorption coefficient or absorber concentration can be calculated. Upon calibration with a gas of known absorption coefficient and knowledge of the concentration of the gas of interest, the ratio of microphone signal to incident radiative power, yields the absorption coefficient. A photoacoustic absorption spectrum is obtained by recording the normalized microphone signal, while varying the wavelength of the incident radiation.

A typical system used for laser excited optoacoustic detection is shown schematically in Figure 2.29. The laser is a CO or CO_2 laser, having a modulated output. The reference detector is used to monitor exciting optical power. Since the laser output power is a function of the wavelength, as well as of operating conditions, the reference detector permits optimization of the

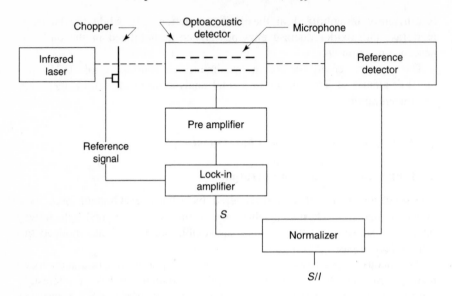

Figure 2.29 Schematic diagram of a typical optoacoustic detection system. Reproduced with permission from Claspy, P.C., et al., *Appl. Optics*, **15**, 1506, © 1976 Optical Society of America.

optoacoustical signals. Phase-sensitive detection of the signal provides noise reduction and thus increase in detection sensitivity.

The transducer used to convert mechanical energy to an electrical voltage output is a condenser microphone, which consists of a thin metal diaphragm or metallized plastic dielectric, and a rigid conducting back plate. When a charge is applied to the capacitor by an external d.c. power supply, capacitance modulation, caused by sound-pressure variation-induced changes in plate separation, will produce a current flow between the plates. The signal output therefore depends on the microphone capacitance, the magnitude of the pressure induced changes in capacitance and the magnitude of the applied voltage. As the signal is inversely proportional to the total circuit capacitance, the microphone capacitance and the amplifier input capacitance, have to be kept as small as possible.

Nearly all gases absorb in the near infrared ($\sim$2–15 μm). For practical considerations measurements are confined to the 5–7 μm range of the continuous wave (CW) CO laser, and to the 9.2–11.5 μm range of the CW CO_2 laser. The selection of the specific spectral region in which to perform the detection is based primarily on the knowledge of the IR spectrum of the molecule of interest. The structures of the absorption spectra of light and heavy molecules are quite different. For light molecules, such as atmospheric pollutants, spectral details are frequently resolved, even at pressures as high as 1 atm, whereas heavy molecules, such as explosives, exhibit unresolved continuum spectra over several wave-numbers in the pressure range. Detection of heavy molecules should therefore use a sequence of laser wavelengths which span the unresolved continuum. For example, the $-NO_2$ group, which is present in most explosives, exhibits an asymmetric stretch vibrational absorption band near 6 μm, and an $-O-N$ stretching mode near 11 μm, with the specific location being determined by the basis molecule to which the group is attached.

2.9.2 Laser optoacoustic detection of explosives

Direct detection of NG, EGDN, and DNT by optoacoustic spectroscopy has been carried out at 6, 9, and 11 μm [84]. The laser was a dry ice cooled, grating tuned, CO laser, which could be tuned from ≤5.8 μm to >6.6 μm. Excitation in the 9 and 11 μm regions was obtained using two grating tuned CO_2 lasers, one using $^{12}C^{16}O_2$ as the lasing gas and the other using $^{13}C^{16}O_2$. All explosives spectra were taken with a 1 mg sample placed inside the cell. The cell was evacuated with the explosive sample in place, after which the cell was filled with dry nitrogen or with air. The cell was heated to 45°C, to reduce surface adsorption. Since a closed system was used, it was assumed that the explosive vapors were in thermal equilibrium with the source. Minimum detectable concentrations were determined by relying on vapor pressure data. Results were 0.28 p.p.b. ($\times 10^{-9}$ g/cm^3) of NG at a wavelength of 11 μm, 1.5 p.p.m. ($\times 10^{-6}$ g/cm^3) of EGDN, also at 11 μm, and 16 p.p.m. of DNT at a wavelength of 9 μm. In the 6 μm region, NO, NO_2, and water vapor gave rise to interference of various degrees, depending on the explosive involved. Of these, water vapor seemed to present the most difficulty. There was no interference in the 9 and 11 μm regions. In another study, detection limits obtained at 9.6 μm (in the absence of interference), were 8.26, 0.23 and 0.5 p.p.b., for EGDN, NG and DNT, respectively [86]. However, ambient levels of CO_2, H_2O (50% relative humidity), NH_3 (1 p.p.m.v.), and O_3 (25 p.p.m.v.) were found to limit detection of EGDN in the p.p.b. range. It was suggested that EGDN could be detected in the presence of ambient levels of these interferrants down to at least 10^{-10}, when excitation is provided by a $^{13}C^{16}O_2$ laser.

2.10 PERSONNEL SCREENING BOOTHS (PORTALS)

During the last 15 years a great effort has been invested in the development of personnel screening systems for detection of concealed explosives. The requirement of such systems is that the sampling and detecting process be performed in very short periods of time, e.g. 6 s. This process consists of four steps.

1. Sampling the air around a person.
2. Capturing the explosives vapor.

3. Transporting the collected vapor to a detection system.
4. Analyzing the collected vapor.

The analysis of the collected vapor is done by one of the methods previously described. This leaves the first three steps for design and improvement. Several booths will be described.

2.10.1 Walk-in/walk-out booth

The booth is designed so that the person enters the booth and walks to the front where he stands facing five sampling ducts, each in the shape of an exponential horn, positioned between the floor and just above the head [87]. The front portion of the booth is a smoothed surface and, because of this, the booth has been described as an 'open clamshell'.

There are also provisions in the booth for warming the person with heat lamps and ruffling the person's clothes with four sets of alternating, high velocity puffers, two directed at the person's front and two directed at the person's back. When the person stands at the sampling location, a large pump is activated and air is exhausted from the booth at 375 L/s for approximately 10 s, so that 3750 L air are exhausted into the sampling ducts for collection and subsequent analysis. This system was able to detect vapors from plastic explosives concealed under a layer of clothing.

Figure 2.30 shows the sketch of a such a portal layout [88]. The booth had a floor, a ceiling and three sides, about 0.75 m wide, 1.25 m deep and 2.5 m high. On the side opposite the open side, there are five intake ports. Each one is a rectangular funnel, connected to a 8 cm pipe. The centers of the five funnels are at 20, 60, 100, 140, and 180 cm from the floor. The five pipes are connected to a mixing box. During sampling, a large blower sucks room air through the open side of the portal around the person, through the five funnels, into the mixing box, and finally to the exhaust. The sample that is used is removed from the mixing box.

2.10.2 Walk-through booth

Walk-through booths are more desirable in high volume situations, such as airport security systems. In this configuration, the person enters the walk-through tunnel and stops at the sampling location, where there are sampling ducts on one side of the person and there may or may not be puffers on the other side of the of the person. There is a sampling pump exhausting air from the booth through the sampling ducts for collection and analysis.

The design of such a booth has to take into consideration an 'air-exchange sweep', sample concentration, temporary sample retention, and purge to the detector. Assuming a required maximum time of 6 s per person, a 200 cubic feet booth with a walk-through transit time of 3 s and an air removal period of 3 s, several compromises have to be considered [89]. One important trade-off

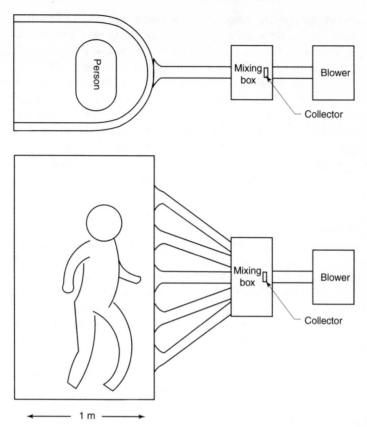

Figure 2.30 Sketch of a portal layout. Reproduced from Bromberg, E.E.A., et al., *Advances in Analysis and Detection of Explosives*, Dordrecht, 1993, p. 473, with kind permission from Kluwer Academic Publishers.

will be the air-velocity at the collection media. For a three-inch diameter collector, the air velocity for the above criteria, would be twice the speed of sound through the preconcentrator, which is, of course, unacceptable. Another consideration is the air collected in the booth with respect to the air loss due to natural walk-through pumping of the transiting individuals. Storage/purge efficiency or desorption of the preconcentration media is of major importance. Budgeting the losses and efficiencies might have the following results:

- 50% of the collected air is lost because of the walk-through pumping.
- 50% of the collected sample is actually deposited on the preconcentration media.
- 80% efficiency of storage and purge to the detector.

Because of these losses, a one femtogram of explosive reaching the detector has to be translated to five femtogram acquired during the booth transit.

In another portal design, mechanical stimulation is achieved by the action of door panels on the subject's clothing, without being intrusive [90]. The level of vapor emissions was increased by the application of infrared irradiation to the subject's clothing and by direct warming of the door panels.

A walk-through portal sampling module that incorporates active sampling has been developed [91]. The portal collects a sample as a passenger walks through an array of wands fitted into the archway. The wands have holes through which the sample is sucked in. The wands brush up against the passenger's body, vacuuming his/her clothing and removing any explosives vapor or particulates that may be present. The sample is pulled into the transport conduit, transported to a collection station, and then filtered from the air stream onto a collector and transferred to the detector for analysis. Since large amounts of air are pulled in through the wands, it is necessary to concentrate the sample by filtering it from the air. The sample collector, or filter, is composed of a porous Teflon membrane with an open area of about 85% and a pore size of approximately 75 μm. Strips of a chemically absorbent polymer on the collector provide for the collection of the more volatile explosives. The collector is desorbed by a heating process. Analysis is done by GC–TEA.

The system uses three sensors. The first one is an ultrasonic proximity sensor, which detects the approach of the passenger, triggers a camera, and takes a picture, which is stored in memory. The second one is a set of height sensors, which will determine the number of open wands according to the passenger's height. The third sensor allows the system to know that the passenger is currently walking through the archway and to continue the sampling process. Following the analysis, the operator is supplied with a 'clear' or 'alarm' signal, indicating whether or not any explosives were found in the sample. The system is capable of sampling passengers at the rate of one every 6 s.

A walk-through portal was designed which uses airflow sampling of explosives vapor, preconcentration, and IMS for analysis [92, 93]. The portal, shown schematically in Figure 2.31, is designed to screen personnel without any direct physical contact. Upon entering the portal, the tested person turns 90° and stands for 5 s, while being screened. Air is blown down from the top of the portal and along the person's body at 7 ft/s, and is sucked into two slots located near the feet. The volume of air collected by each of these slots is 160 L/s. The airflow sampling is designed to collect both vapor and particulates, if present on a person's clothes. The total screening time per person tested is about 12 s. After passing through the slot, the air flows into a preconcentrator (a molecular filter which collects the explosive molecules onto a screen) concentrating about 800 L air into a 2 L sample. When the air collection cycle is finished, this screen is heated to desorb the collected explosives back into the gas phase, and the resulting explosives-enriched vapor is pulsed into an IMS for detection.

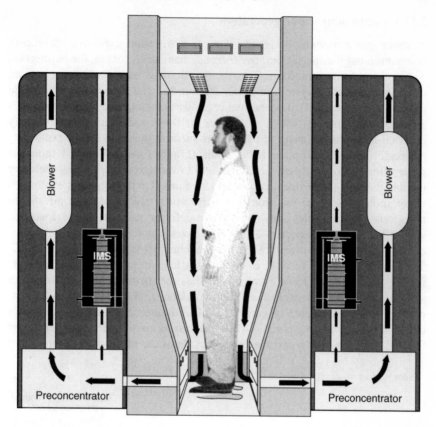

Figure 2.31 Explosives detection portal. Reproduced from Parmeter, J.E., et al., Second Explosives Detection Technology Symposium, Atlantic City, NJ, 1996, p. 187, with permission.

Testing of the portal was carried out with TNT and with RDX. It was found that the system response to TNT vapor was linear in the range 0–300 ng: 100 ng of TNT could easily be detected.

Testing of the portal was also performed with cloth patches containing known amounts of C-4 (containing 94% RDX) particulate, which were attached to a person, walking into the portal. The patches contained 3–9 µg of C-4. Positive detections were made in all cases, with the amount of RDX reaching the IMS being 0.1–1% of the original amount present in the patch.

2.11 EXPLOSIVE VAPOR GENERATORS

A key element in the testing and evaluation of explosives vapor detectors is a calibrated vapor source or vapor generator. Several vapor generators will be described.

2.11.1 Continuous vapor generators

A vapor generator was designed, using a permeation explosives source (a permeation bag), a special-purpose thermal chamber to contain the permeation bag, and a downstream dilution device that delivers explosives vapor masses to a vapor detection system in an air stream that can be varied over a wide range of flow rates [94]. The thermal chamber is maintained at 75°C and uses a porous wall, through which the carrier air flows and forms a continuously renewed air boundary layer at the porous surface, thus preventing explosives molecule surface adsorption to its walls (Figure 2.32). A filtered room air stream flows into the thermal chamber at 1 L/min and to the dilution-delivery module through a vent-line scrubber and is returned to the room. As explosives vapor molecules come out from the permeation bag, the carrier air stream transports the molecules into the dilution-delivery module, where a second stage dilution process operates on the explosives vapor molecules. All elements of the generator downstream of the thermal chamber are maintained at 150°C. The dilution-delivery module is basically a temperature-controlled semiper-meable membrane at 150°C, which separates the thermal chamber air carrier stream containing the explosives vapor molecules and the dilution-delivery module carrier air stream (Figure 2.33). The dilution room air stream passes over one side of the membrane, while the carrier air stream coming from the thermal chamber, flows past the opposite side of the membrane. A small part

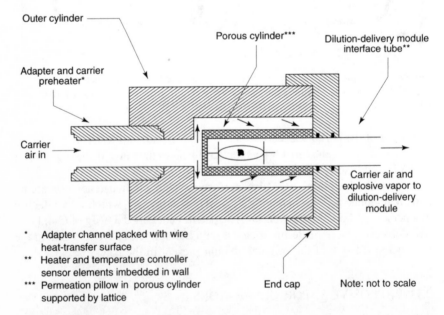

Figure 2.32 Vapor generator thermal chamber. Reproduced from Lucero, D.P., et al., *Advances in Analysis and Detection of Explosives*, Dordrecht, 1993, p. 485, with kind permission from Kluwer Academic Publishers.

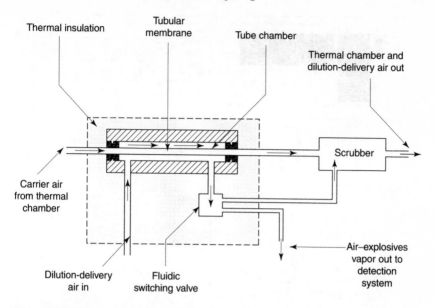

Figure 2.33 Vapor generator dilution-delivery module. Reproduced from Lucero, D.P., et al., *Advances in Analysis and Detection of Explosives*, Dordrecht, 1993, p. 485 with kind permission from Kluwer Academic Publishers.

of the explosive vapors passes through the membrane by a diffusion process, into the dilution-delivery module carrier air stream at a rate proportional to the product of the membrane permeation conductance and the explosives vapor partial pressure, or concentration, of the thermal chamber carrier stream. Thus, two stages, the permeation bag and the membrane, are used to produce an air stream containing controlled and ultra-low-level concentrations of explosives vapor. For example, the generator operating with an RDX permeation bag and a dilution-delivery module, using 1 L/min flow rate, yields an output concentration of 2×10^{-3} p.p.t. or 2×10^{-14} g/min RDX mass flow rate. A vapor detection system ingesting the output carrier air stream for 3 s will absorb 1×10^{-15} g of RDX.

The permeation bag explosives vapor source is a permeation device that outgasses explosives vapor molecules at a prescribed rate at a given temperature. It is maintained at 75°C to obtain a sufficiently high vapor emission rate. A typical permeation bag consists of a thin, flat film of Teflon, folded over itself, heatsealed around its edges, and containing explosives material. For example, a typical $60 \, \text{cm}^2$ bag of 0.001 inch thick Teflon film, containing 0.5 g of RDX, will produce a RDX vapor output of 5.3×10^{-11} g/min or 0.53 µg/week at 75°C. This figure is based on a Teflon permeability of 3.8×10^{-8} std mL/(min-cm^2-torr/cm) for all explosives. The dilution-delivery module further reduces the vapor emission rate by several orders of magnitude, in a precise and controlled way by adjusting the air flow rate.

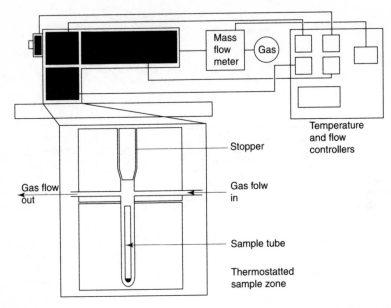

Figure 2.34 Schematic diagram of explosives vapor generator. Reprinted from Eiceman, G.A., et al., *Talanta*, **45**, 57, © 1997, with permission from Elsevier Science.

Another vapor generator, shown in Figure 2.34, consists of a sample container, a heated block to thermostat the sample container, temperature controllers for all surfaces of the vapor generator, and a gas flow controller [95]. The sample container was a precision-made glass tube, 0.47 cm inner diameter × 10 cm length. Nitrogen flow rate was 130 mL/min. The four regions of the generator, equipped with individual temperature controllers, included inlet, sample container, diffusion, and outlet regions.

In one configuration, the generator outlet flow was directed fully into an atmospheric pressure chemical ionization (APCI) tandem mass spectrometer (MS/MS), a fume hood, or a cold trap. In a second configuration, called 'discrete mode of sampling', a switch box was added to the outlet flow, so that the vapor stream could be diverted, in a controlled and timed manner, into an ion mobility spectrometer (IMS). Any excess flow was vented to a fume hood. A tandem mass spectrometer was coupled to an IMS for chemical characterization of the gas stream.

Small amounts of TNT, RDX, or PETN were placed into a sample tube and weighed on an analytical scale: 10–20 mg of sample were placed in the sample tube, which was inserted into the vapor generator through an access port, which was then closed with a glass stopper. The vapor generator was operated continuously with pre-set gas flows and temperatures. Gravimetric measurements of the sample and container were made at time intervals after the introduction of the sample into the generator. The generator was operated for 24–48 h with a sample before the first mass determination was made, to allow for equilibration of all surfaces where sample vapors flowed and removal of highly volatile impurities, if present, either in the sample or sample tube.

Output in mass as a function of time was measured for explosives in the vapor generator at specific temperatures in the range 79–150°C. Figure 2.35 shows the mass output results for TNT. The rates varied from 84 pg/s at 79°C to 13 ng/s at 150°C. The generator output of TNT vapor was stable over 400 h of continuous operation. An independent calibration of the vapor generator using an IMS to measure the mass output gave a rate of 85 pg/s at 79°C, in good agreement with the gravimetric results.

Mass loss measurements of RDX also indicated a stable mass output rate for at least 300 h of continuous operation, with output rate varying from 49 pg/s at 110°C to 2.4 ng/s at 150°C. Measurements with PETN showed that the vapor output was not pure PETN, but also contained a large amount of decomposition products.

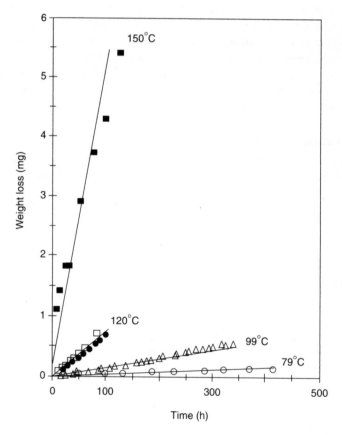

Figure 2.35 Vapor generator mass output of TNT at different temperatures. Reprinted from Eiceman, G.A., et al., *Talanta*, **45**, 57, © 1997, with permission from Elsevier Science.

2.11.2 Transient vapor generator

A gas chromatograph (GC) equipped with a capillary split/splitless injector and an electron-capture detector (ECD) was converted for use in this generator [96]. The capillary column extended outside the GC by 9.5 cm. A special heater was designed for the exposed part of the capillary column. The probe of the detection system was placed 1–2 mm from the end of the capillary column. To prevent air currents from interfering with detection of the explosives vapors, a piece of quartz tubing was placed around the capillary column heater and probe, as an air shield (Figure 2.36). The calibration of the vapor generator was carried out by the injection of known standards of highly pure explosive solutions into the GC with the capillary connected to the ECD. To test an explosive vapor detection system, the fused silica capillary was removed from the ECD and placed in the heater system, close to the probe of the detection system.

The use of a GC provides pure vapor, because all volatile impurities will be separated and elute at different times. The explosives elute from the vapor generator as a transient peak. Since the vapor cloud of explosive is present for less than a minute, there is neither enough time nor enough explosive to saturate a detection system's concentrator and permit the explosive to pass through the detector. The transient nature of the vapor produced, permits detection systems to be tested in a manner which evaluates the abilities of the system to capture, analyze, and respond to specific levels of explosive vapors.

2.11.3 Pulsed vapor generators

A generator, based on the sampling preconcentrating collector [42], was developed for the evaluation of portal detection devices [88]. It uses a collector to

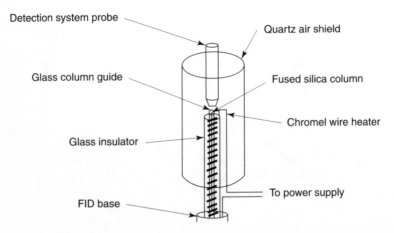

Figure 2.36 Effluent heater system in transient vapor generator. Reproduced from Reiner, G.A., et al., *J. Energetic Mater.*, **9**, 173, (1991), by permission of Dowden, Brodman and Devine, Inc.

preconcentrate the sample. The collector consists of a metal ribbon that is wound around a hub. The sample is sucked through the collector, and a fraction of the vapors and particles stick to the surface. The ribbon collector is heated rapidly to the desired temperature, by a current pulse passing through it. The explosive molecules that have accumulated on the ribbon surface vaporize and are transferred to the detector. To calibrate the device, a known amount of different explosive mixtures in solution is injected onto the collector. The collector is then connected to the detector, fired, and the vapors analyzed by GC–TEA. It was found that over 90% of the explosives injected onto the collector were desorbed during the first 2 s of firing.

Another pulsed explosives vapor generator was constructed for quantitation of vapor standards (Figure 2.37) [97, 98]. The system is composed of four parts: a temperature-controlled explosive vapor reservoir, a flow control manifold, a supply of temperature-controlled air, and a data control and acquisition system. The explosive reservoir contains about 0.1 g of solid explosive suspended on quartz beads. The explosive is dissolved in ≈15 mL of methyl ethyl ketone. About 20–22 g of 0.1 mm diameter quartz beads are added to the solution. The explosive-coated beads are then loaded into a 10 ft long thin walled 1/8 inch outer diameter stainless steel tube. Quartz wool plugs are

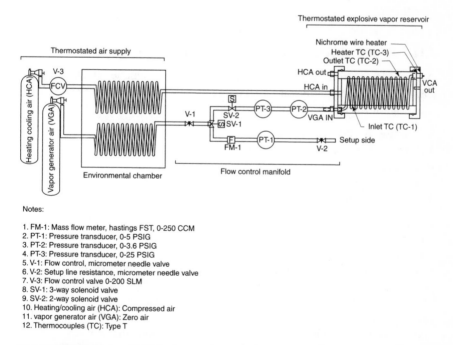

Notes:

1. FM-1: Mass flow meter, hastings FST, 0-250 CCM
2. PT-1: Pressure transducer, 0-5 PSIG
3. PT-2: Pressure transducer, 0-3.6 PSIG
4. PT-3: Pressure transducer, 0-25 PSIG
5. V-1: Flow control, micrometer needle valve
6. V-2: Setup line resistance, micrometer needle valve
7. V-3: Flow control valve 0-200 SLM
8. SV-1: 3-way solenoid valve
9. SV-2: 2-way solenoid valve
10. Heating/cooling air (HCA): Compressed air
11. vapor generator air (VGA): Zero air
12. Thermocouples (TC): Type T

Figure 2.37 Pulsed explosives vapor generator. Reproduced from Davies, J.P., et al., *Advances in Analysis and Detection of Explosives*, Dordrecht, 1993, p. 513, with kind permission from Kluwer Academic Publishers.

packed into both ends of the tubing to hold the beads in place. The tube was coiled around a 2 inch aluminum tube and placed into a temperature-controlled chamber. Ultra-pure air was used as carrier gas and passed through the coil to carry the explosive molecules. The generator is capable of delivering a pulse of explosive vapor, by means of a pulse of air flow through the coil, which is controlled by temperature, air flow rate and pulse width. During the initial setup and between explosive vapor pulses, the carrier gas is diverted using a three-way solenoid valve (SV-1) to the setup manifold. The setup manifold is used to simulate the resistance of the generator by adjusting a needle valve (V-2) at the end of the manifold. A preconcentrator was used to collect the explosive vapors from the vapor generator for subsequent quantification by IMS. The typical sampling method includes a 5 s presampling time, followed by 5 s sample pulse and a 2 s post-sampling time. After the 12 s the preconcentrator is removed from the vapor generator outlet and placed into the IMS inlet, where a heater will desorb the explosives into the IMS.

Three such vapor generators were constructed, one for each one of the explosives TNT, RDX, and PETN. The generators were calibrated in the following ranges: 52–445 pg for TNT, 109–1000 pg for RDX, and 326–1000 pg for PETN. The vapor generators were also checked with a triple quadrupole MS/MS mass spectrometer [99]. IMS and MS/MS data were in good agreement.

2.11.4 Canine olfaction vapor delivery system (olfactometer)

A quantitative vapor delivery system for the study of canine olfaction and the training of dogs on particular target odorants, has been constructed [100, 101]. The criteria for the design of the olfactometer were as follows:

- The use of odor-free purified air.
- Temperature and humidity control.
- Control of odorant concentration.
- The use of odorless components.
- The use of pure stimulus compounds.

In addition, the design took into consideration issues of contamination such as chemical adsorption and cross-contamination between delivery lines or separate odorants.

The olfactometer, based on these criteria, was a multidilution odor delivery system, consisting of concentrated odorant streams, diluted with purified air at preselected flow ratio levels. The olfactometer was heated to 31°C during operation, while the odorant source was kept in a water bath at 27°C. The odorant sample vessels (Figure 2.38) were made of borosilicate glass. A fritted glass disc was fused 0.5 in from the bottom of the vessel. Purified air was introduced into the vessel and forced up through the fritted glass disc. The air then flowed through the odorant source to produce a concentrated vapor

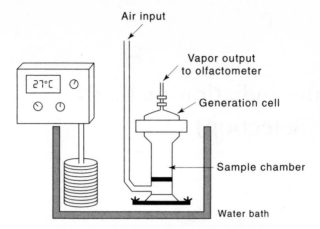

Figure 2.38 Olfactometer vapor generation cell. Reproduced from Hartell, M.G., et al., 5th Int. Symp. on Analysis and Detection of Explosives, Washington, DC, 1995, with permission.

which entered the olfactometer. Computer-controlled solenoid valves within the olfactometer allowed automated operation.

The odorant source used for the testing of the system was Hercules Unique smokeless powder, a nitroglycerin/nitrocellulose double-based powder. A sample of 30 g of powder was placed in the vapor generating vessel. Analysis was performed by GC/MS.

Standards used for quantitation included a solution containing hydrocarbons (alkanes C_{10}–C_{20}), and a solution containing NG, EGDN, and 1,2-propanediol dinitrate.

3

Probing radiation methods (bulk detection)

3.1 INTRODUCTION

Because explosives can be hidden inside various types of luggage, containing many different materials, effective radiation methods are limited to penetrating electromagnetic radiation, low-energy radiation radiowaves and microwaves, high-energy X-rays and γ-rays and neutrons. Although radiation methods that depend on compound-specific interactions, such as infrared, visible light and UV radiation, would allow the differentiation of organic explosives from other materials, these methods cannot be used because they do not penetrate the container examined.

This chapter is arranged according to the type of radiation incident on the object examined.

3.2 PROPERTIES OF EXPLOSIVES

It is of interest to look at some of the properties of explosives with regard to nonintrusive radiation methods for their detection.

Explosives are composed of hydrogen, carbon, nitrogen, and oxygen, and many other organic compounds. Explosives, as a group, are rich in nitrogen and oxygen, poor in carbon and hydrogen [102, 103]. Atomic concentration is not a good discriminant, since many common nonexplosive compounds have high atomic concentrations of oxygen (Dacron, silk, cotton) and nitrogen (wool, silk, Orlon). A better discriminant is the atomic density (moles per cm^3) of oxygen and nitrogen. Still, many common materials, such as water and sand, have high atomic densities of oxygen, but very few have high densities of nitrogen. Exceptions are Melamine, polyurethene and solid nylon. Figure 3.1 shows the nitrogen density of common materials and explosives [104, 105]. Figures 3.2–3.4 show the atomic densities of carbon and hydrogen, carbon and oxygen, and nitrogen and oxygen, respectively, of explosives and

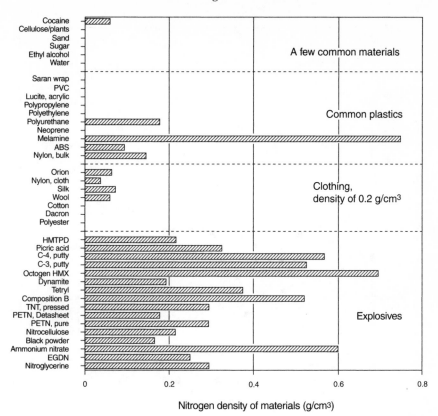

Figure 3.1 Nitrogen density of common materials and explosives. Reproduced from Reports OTA-ISC-481 (1991) and OTA-ISC-511 (1992), US Congress, Office of Technology Assessment, Washington, DC.

common nonexplosive compounds [102]. A measurement of the oxygen and nitrogen densities, to an uncertainty of 20%, gives a unique separation of explosives from other compounds.

3.3 ELECTROMAGNETIC RADIATION

3.3.1 X-Rays

When a rapidly moving electron impinges on an atom, it may knock an electron completely out of one of the inner orbits of that atom. When electrons from the outer orbits fall to the vacant inner orbit, the difference in energy is emitted as electromagnetic radiation, called X-rays.

The X-rays useful for detection of explosives have energies from a few thousand up to several million electron volts, or wavelengths in the range

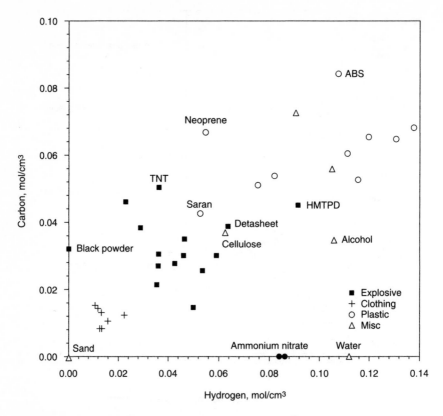

Figure 3.2 Atomic densities of carbon and hydrogen of explosives and common nonexplosive compounds. Reproduced from Grodzins, L., First Int. Symp. on Explosive Detection Technology, Atlantic City, NJ, 1991, p. 201, with permission.

10^{-8}–10^{-11} cm. Although there is no precise definition of the high-energy limit of the energy of X-ray photons, this term is usually restricted to radiation having an energy below several MeV, above which the radiation is referred to as gamma (γ) radiation [106]. There is a tendency to distinguish between X-rays and γ-rays on the basis of their mode of generation, calling them X-rays if generated by processes occurring outside the atomic nucleus, and γ-rays if generated within an atomic nucleus.

3.3.1.1 Interaction of X-ray photons with matter

The absorption of a beam of X-rays of energy E passing through a material, can be described by the following equation:

$$I = I_0 e^{-\mu l \rho} \qquad (3.1)$$

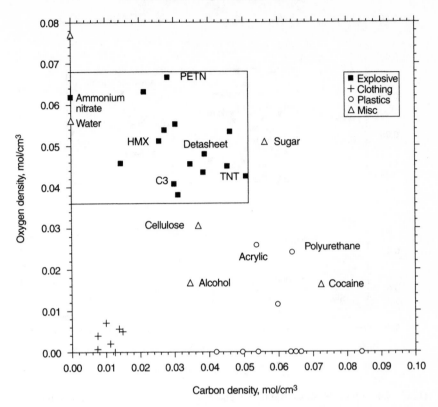

Figure 3.3 Atomic densities of carbon and oxygen of explosives and common nonexplosive compounds. Reproduced from Grodzins, L., First Int. Symp. on Explosive Detection Technology, Atlantic City, NJ, 1991, p. 201, with permission.

Where: I = intensity of emergent beam (photons/s)

I_0 = intensity of incident beam

μ = total mass attenuation coefficient, describing both absorption and scattering (cm^2/g)

l = length of path through the absorbing material (cm)

ρ = density of absorbing material (g/cm^3)

μ depends upon the energy of the X-rays and upon the effective atomic number Z_{eff} of the absorbing material. As the energy of the X-rays increases, μ decreases. Z_{eff} is equivalent to the average number of electrons per atom.

X-ray photons are removed from the beam in one of three ways:

- Photoelectric absorption: an X-ray photon hits a bounded electron in an atom, transferring its energy to it, and knocking it out of the atom.
- Compton scattering: an X-ray photon interacts with an electron and transfers part of its energy to the electron. The electron is knocked out of the atom

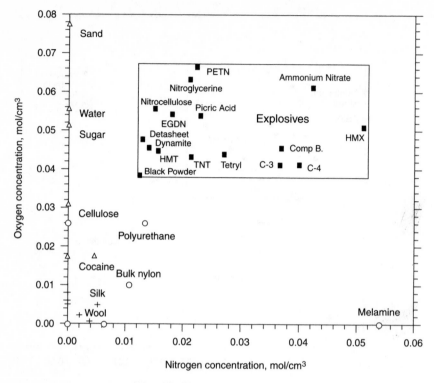

Figure 3.4 Atomic densities of nitrogen and oxygen of explosives and common nonexplosive compounds. Reproduced from Grodzins, L., First Int. Symp. on Explosive Detection Technology, Atlantic City, NJ, 1991, p. 201, with permission.

and a new photon of lower energy proceeds from the collision, in an altered direction.

- Pair production of a positron and an electron results when a high-energy X-ray photon is annihilated following interaction with the nucleus of a heavy atom. This occurs at X-ray energies above 1.022 MeV.

The photoelectric effect increases as the fifth power of the atomic number and decreases as the 7/2 power of the photon energy. For metals, such as iron, the photoelectric effect is dominant up to energies of 100–150 keV, whereas for organic compounds, such as plastic explosives, the Compton effect dominates at energies above 20 keV.

3.3.1.2 Single view, single energy X-ray system

The common airport X-ray scanner operates with electron energies of 120 keV impinging on a tungsten target. The resulting beam has the characteristic energy of tungsten ($\approx$60 keV). Airport X-ray imaging systems were designed to give high resolution pictures with excellent gray scale dynamic range so that the operator could identify weapons made of metal and having characteristic shapes. We now ask these machines to detect explosives, made from light materials with no characteristic shape.

The attenuated signal is supposed to be a measure of a potential threat. The total signal from a piece of luggage is the sum of the signals from every component in the bag. As suitcases contain a great variety of items, light material as well as heavy materials, such as metals, we will obtain a false alarm on almost every tested item. Explosives can be placed behind or within items of higher atomic number and will remain undetected.

3.3.1.3 Dual-energy X-ray system

Dual-energy analysis makes use of the fact that the photoelectric effect is strongly dependent on the atomic number. Each energy generates a different mass attenuation coefficient, μ. Thus, by comparing the transmission from two images produced by different beam energies, information about the effective atomic number, Z_{eff}, of the examined object can be obtained. Because the energy spectra are broad and not providing enough distinct image information, the determination of Z_{eff} by this method is only approximate. The system distinguishes between higher and lower atomic number along the beam in the tested object. For example, it can distinguish dense organic items from metals and from less dense organic items, such as food, clothes, and other items usually found in suitcases. Figure 3.5 shows the ratio of mass attenuation coefficients for a number of materials and for two energy ratios: 100/60 keV and 80/40 keV [102]. It can be seen that the lower the X-ray energy, the better the discrimination power. However, lower energy X-ray photons are strongly absorbed and it is therefore assumed that present X-ray systems will not be able to produce strong enough signals with 40 keV X-rays passing through heavy luggage.

3.3.1.4 Compton scattering

When an incident X-ray photon undergoes a Compton scattering interaction, there is a scattered photon whose energy is a function of the scattered angle as follows [106]:

$$\lambda = \lambda_0 + \lambda_1(1 - \cos\alpha) \tag{3.2}$$

where: λ = wavelength of the scattered photon
λ_0 = wavelength of incident photon
$\lambda_1 = h/mc$
h = Planck's constant
m = mass of electron
c = speed of light
α = scattering angle of the photon

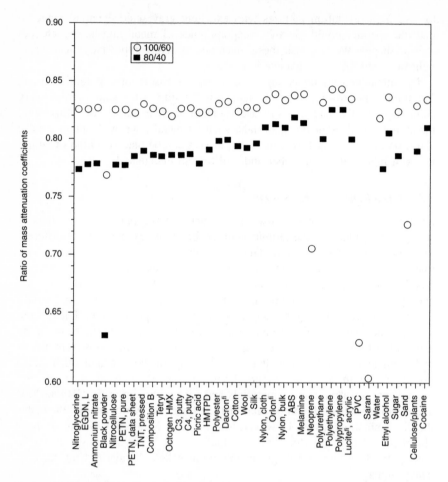

Figure 3.5 Ratio of mass attenuation coefficients for various materials and for two X-ray energy ratios: 100/60 keV and 80/40 keV. Reproduced from Grodzins, L., First Int. Symp. on Explosive Detection Technology, Atlantic City, NJ, 1991, p. 201, with permission.

The energy of this scattered photon is reduced by the recoil kinetic energy of the struck electron. For 100 keV X-rays, the back-scattered photon is reduced in energy by nearly 40%, increasing significantly the mass attenuation coefficient, μ. Photons scattered at large angles are thus even more strongly absorbed by the photoelectric effect on the way out of the tested item, then are unscattered photons that make up the transmitted beam. These back-scattered photons can be measured and, together with the absorption measurement, provide information about the tested item. This information can separate the effects of density and atomic number, allowing the system to indicate high-density, low Z material, the signature of explosives. Figure 3.6 shows a schematic

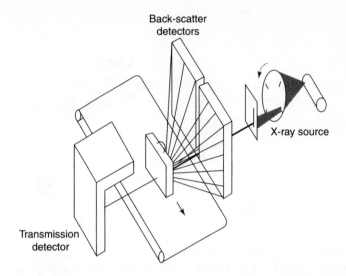

Back-scatter
detectors

X-ray source

Transmission
detector

Figure 3.6 Schematic diagram of transmission and back-scatter detection system. Reproduced from Schafer, D., et al., First Int. Symp. on Detection Technology, Atlantic City, NJ, 1991, p. 269, with permission.

diagram of such a system [107]. Compton back-scattering gives information about light material in the near surfaces of luggage, because Compton signals from heavier material are attenuated by photoelectric absorption. Because the signals from the heavier materials are suppressed, the light materials stand out clearly. By proper choice of the beam conditions, the light materials in the near surfaces of luggage (and to a lesser degree inside) can be strongly enhanced for either visual or automated discrimination.

A dual energy detection system with two scattering detectors (Figure 3.7), was built to improve sheet explosive threat identification [108]. The first is a forward scattering detector, located close to the surface of the inspected bag, from which the transmitted X-rays are exiting. The X-rays observed are scattered at angles less than 90°. The second detector is a backward scattering detector, located near the surface of the bag where the X-ray beam enters. Scattering angles for this detector are greater than 90°. The physics of forward and backward scattering are basically the same, but there are important measurement differences due to the geometry. At the X-rays energies used, 20–150 keV, the Compton scattering cross section depends on the electron density of the scattering material, and therefore varies little with the atomic composition of the material. The most significant factors affecting the intensity of the detected X-ray beam are the thickness and density of the target material.

The forward scattering detector is an array of photomultiplier crystal X-ray detectors placed near the X-ray fan beam, just beyond the object being inspected, as shown in Figure 3.7. Collimation and shielding are arranged

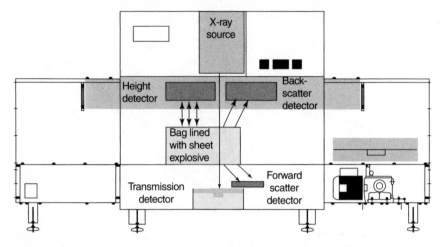

Figure 3.7 Dual energy detection system with two scattering detectors. Reproduced from Aitkenhead, W. F., et al., Second Explosives Detection Technology Symposium, Atlantic City, NJ, 1996, p. 236, with permission.

around the detector elements so that only X-rays scattered in the lowest 3 cm of the inspected bag may be seen by the array.

While the source is emitting X-rays, data is acquired from both the forward scattering detector and the transmission detector, measuring both the scatter signal from the target layer and the total transmitted signal.

The back-scattering detector is also an array of photomultiplier crystal assemblies, but they are placed near the X-ray fan beam on the source side of the inspected bag. Shielding and collimation are arranged so that only X-rays generated from scattering within the inspected bag and along each pixel are detected.

Since the height of the inspected object can vary, whereas the detector array is fixed, scattering from any depth in the bag will be detected. However, we need to know the distance from each scattering target location to the scattering detector array. The scattering signal of an object with and without a sheet explosive threat can be evaluated if we define an acceptance function that is a measure of how many X-ray photons are detected by the scattering detector as function of the height of the target layer.

For example, we add to the top of a bag, with a uniform material distribution of $0.2\,g/cm^3$ (which is representative of a bag packed with clothing) a sheet explosive. This will change the material distribution of the bag, since sheet explosive has a high density of about $1.5\,g/cm^3$. Back-scattering data was found to have almost 50% more contrast than the transmission data.

An infrared (IR) height detector was incorporated into the system to make a pixel by pixel determination of baggage height. This was found to be very helpful in finding target objects. The baggage height is determined by measuring the reflection angle of each IR beam striking the top surface of the bag. Dual energy data from all three detector arrays (front-scattering, back-scattering, and height detectors) are collected synchronously and are correlated with transmission data on a line by line basis. The scattering data are normalized on a pixel by pixel basis for multiplier gain, detector geometry, the bag height profile, and the attenuation characteristics of the bag, as measured by the transmission detectors. Algorithms have been developed and tested that find sheet explosives in bags and generate automated alarms, indicating the location of the threat [108]. There are multiple detection algorithms because there are several ways that a sheet explosive could be placed in a suitcase, each producing different signatures in different images. The detection algorithms are designed to examine any possible placement of a sheet explosive in a bag.

3.3.1.5 Low-angle X-ray scattering (LAXS)

The application of low angle coherent scattering of X-rays has been proposed for the detection of sheet explosive, concealed in passenger luggage [109]. An experimental system has been developed to measure the energy distribution of photons scattered at a well-defined angle from within a suitcase. The energy distribution of the incident photons is modified by the diffraction effects, characteristic of coherent scatter, and results in a unique pattern, which depends on the molecular structure of the scattering medium. Thus, the explosive will be differentiated from benign materials, normally present in luggage.

Many explosives have a crystalline structure, and therefore produce characteristic diffraction effects when irradiated by photons. This phenomenon is exploited by the LAXS method, where a measurement is made of the energy distribution of photons scattered at a well-defined angle (typically $<10°$). The experimental arrangement is shown schematically in Figure 3.8. The source is a tungsten target industrial X-ray tube, operating in the range 0–160 kV and 0–30 mA. The tube has 1 mm Be inherent filtration and is operated in broad focus with a 3×3 mm focal spot. The detector is a planar HPGe detector with crystal dimensions of 16 mm diameter and 10 mm thick. The primary source has a circular 3 mm diameter exit aperture. The incident and scattered beam collimation is formed by interlocking leaf collimators, 50 mm in height and of variable slit width. The incident beam collimation forms a diverging ribbon beam of 16 mm height, at the sample. The detector collimation can be oriented at any selected angle, relative to the incident beam. The width varied according to the required angular resolution. The scatter angle, θ, is defined as the angle subtended at the intersection of the central axis of the primary and scattered beam. The slit collimation can be replaced by a pin-hole arrangement to convert the ribbon beam geometry to a pencil beam geometry.

Two explosives were used in this study: PE4, containing approximately 95% RDX and 5% oil-based binding material, and Semtex, containing approximately 45% RDX, 45% PETN, and 10% oil-based binder. The samples

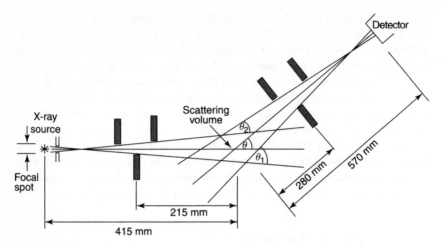

Figure 3.8 Low-angle X-ray scattering (LAXS) system. Reprinted from Luggar, R. D., et al., *Appl. Radiat. Isot.* **48**, 215, © 1997, with permission from Elsevier Science.

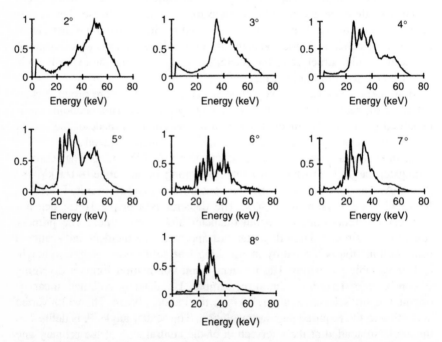

Figure 3.9 PE4 diffraction profiles at scattering angles of 2° to 8°. Reprinted from Luggar, R. D., et al., *Appl. Radiat. Isot.*, **48**, 215, © 1997, with permission from Elsevier Science.

were formed into 5 mm thick sheets. If the minimal amount of explosive which has to be detected is 250 g, with a typical density of 1.8 g/cm³ and a thickness of 5 mm, an area of approximately 280 cm² would result.

The choice of the scattering angle is very important, because it will affect the resolution, the energy and the intensity of the diffraction profiles. Figures 3.9 and 3.10 show the profiles obtained from PE4 and Semtex, respectively, measured at a range of 2° to 8°, at 1° intervals. Measurements were made at 70 kV with wide slit collimators and an acquisition time of 100 s, to ensure good statistical reliability.

The fundamental relationship between angle and energy for diffraction, as derived from Bragg's law is:

$$E^{-1} \propto \sin(\theta/2) \tag{3.3}$$

where θ is the angle through which a photon of energy E is scattered. From Eqn 3.3 it can be seen that the angle at which a photon is scattered from a given scattering center is inversely proportional to its energy. The region of momentum interrogated depends upon the scattering angle: the larger the scatter angle θ, the larger the momentum region, and hence the more compressed the diffraction profile, which means more information within a

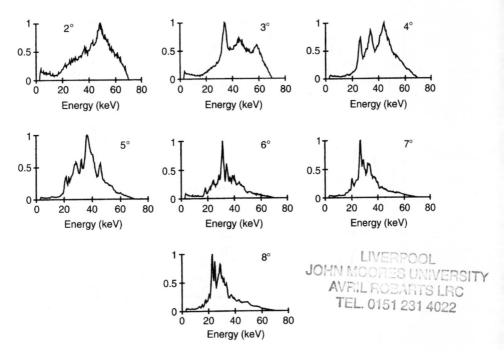

Figure 3.10 Semtex diffraction profiles at scattering angles of 2° to 8°. Reprinted from Luggar, R. D., et al., *Appl. Radiat. Isot.*, **48**, 215, © 1997, with permission from Elsevier Science.

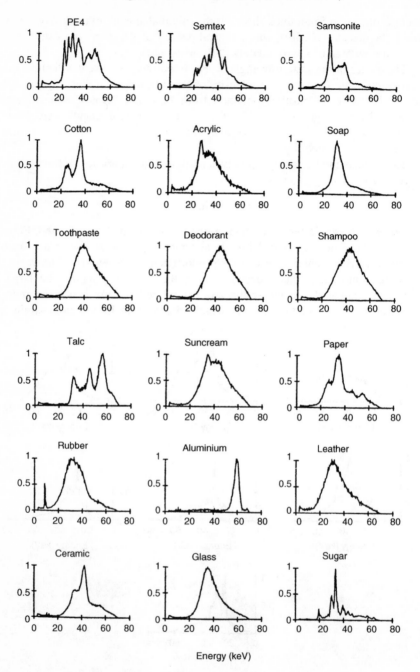

Figure 3.11 Diffraction profiles of typical suitcase contents. Reprinted with permission from Luggar, R. D., et al., *Appl. Radiat. Isot.*, **48**, 215, © 1997, with permission from Elsevier Science.

defined energy width. An additional point of interest is that the higher the scattering angle, and therefore the lower the energy of the diffraction peaks, the greater the resolution of the diffraction profile for a given collimation width.

If the analysis was limited to single samples of thin materials, then the best geometry would be to use large scattering angles to investigate a large momentum region and produce highly resolved characteristic diffraction peaks at low energy. However, in this application, thick items, such as suitcases containing a variety of materials, are used. In order to penetrate such items, the structure in the diffraction profile must occur at a sufficient high energy. The lowest energy diffraction peaks should occur at no less than 20 keV, and ideally should be much higher.

The two most significant factors in selecting the best geometry are the need for good energy resolution and high photon energy. These are contradictory parameters, requiring high scattering angles and low scattering angles, respectively. An experimental compromise is a scattering angle of 4–5°, which, together with an angular resolution of about 20%, will yield well-resolved diffraction profiles, suitable for detection of explosives in passenger luggage. An investigation of the diffraction properties of benign materials, present in passenger luggage, has been carried out and is summarized in Figure 3.11. Many of the materials produce broad featureless diffraction profiles, with a single maximum at approximately 40 keV. This is indicative of low-order molecular structure and may easily be differentiated from the explosives. There are several exceptions, such as cotton, talc, and paper, which produce characteristic diffraction patterns, although different from those of the explosives. Sugar is the only one of the benign materials tested, that produces more than three discrete diffraction peaks. Suitable pattern recognition analysis could pick out individual explosives from the benign materials.

A similar method, called angular dispersive X-ray diffraction (ADXRD) has been applied to the detection of explosives [110]. A custom designed collimator was used to obtain diffraction profiles from independent voxels within luggage. The prototype results from the explosives Semtex, RDX, and PETN showed unique diffraction profiles, which could be differentiated from those of benign materials. Angular- and energy-resolved diffraction profiles were acquired in order to develop data analysis algorithms and make quantitative predictions of the detection probability.

Another method, based on X-ray diffraction, called coherent X-ray scattering (CXRS), has been applied for the detection of explosives in airport luggage [111]. This method is based on another tomographic imaging technique, called energy-dispersive X-ray diffraction tomography (EXDT) [112]. This technique is based on the energy analysis, at fixed angle, of coherent

X-ray scatter excited in an object irradiated by polychromatic radiation. It allows measurement of the X-ray diffraction pattern from a small voxel within an extended object. Tomographic information is obtained directly, without the need to reconstruct from projections.

In CXRS, a primary collimator provides a specially pencil-shaped primary X-ray beam, which scans the suitcase. The coherent scatter radiation of ten adjacent volume voxels in the vertical direction is diverted, by a second collimator, to a segmented eleven-channel germanium detector. A single detector is thus able to produce signals of ten scattering voxels as well as the transmission signal. Apart from the scatter angle, the collimator defines the position and volume of the scatter voxel within the object. The energy distribution of the scattered X-rays is obtained by a pulse height analysis of the detector signals by the data acquisition system. The scattering geometry, which requires small scattering angles to focus the elastic scatter radiation, has been optimized, so that bulk objects can be interrogated with relevant photons having energies above 40 keV. It was found that most of the explosives, due to their polycrystalline nature, exhibit characteristic peaks, with well defined positions and heights, and could therefore, be distinguished from one another and from benign nonexplosive materials [111]. A prototype detector was built, consisting of an X-ray system with a high-power 160 kV rotating anode tube, 5 kW continuous load, 10 kW load for a period of 5 min. The scan time needed for a piece of luggage of medium size was approximately 90 s, with a tube power of 10 kW. The minimum amount of explosive detected was 200 g. Sheet explosives (5 mm thick) could also be detected.

3.3.1.6 High-energy X-ray imaging

Systems using X-rays of 80–100 keV are marginally capable of examining the bulkier pieces of checked luggage. As the energy of the X-ray beam is increased in order to increase the penetrating power, the interactions become dominated by forward Compton interactions which are independent of Z [102].

In high-energy systems, e.g. 10 MeV, pair production cross section is important. Such systems can be useful in scanning large containers. They use beams from linear accelerators and they are very large, as they need appropriate shielding. In order to obtain information on the atomic number of the absorbing material, one needs two measurements, taken at two different energies, e.g. 5 and 10 MeV. It has been found that the discriminating power at energies in the MeV range, is at least as great as that in the 50–100 keV range.

3.3.1.7 Computed tomography (CT)

Computed tomography (CT) is the numerical reconstruction of a cross-sectional image from X-ray projections at various angles around an object [113]. The CT image formed is a mapping of X-ray attenuation properties (both absorption and scattering) of each volume element within the cross section, which can be correlated with the object's density and composition. The projections are formed by scanning a thin cross section of the object with a narrow X-ray beam and measuring the transmitted radiation. The detector does not form the image. It only adds up the energy of all the transmitted photons. The numerical data from multiple ray sums are then computer-processed to reconstruct an image. Image quality is related to the number of ray projections used to reconstruct each CT scan image.

Resolution is composed of two components: spatial resolution and contrast resolution. Spatial resolution of a CT unit is the ability to display, as separate images, two objects that are very close to each other. Contrast resolution of a CT unit is the ability to display, as distinct images, areas that differ in density by a small amount. Contrast and spatial resolution are intimately related to each other and to the amount of radiation absorbed by the detector.

The CT explosives detector produces both projection X-ray and CT images [114]. The detector rotates the X-ray source and detector array around an object to create cross-sectional images of the luggage tested. These cross-sectional images are commonly known as 'slices'. Data from each CT slice are rapidly acquired and processed by a computer. The data gathered from the slices allows the system to measure accurately the physical characteristics of each object, regardless of its shape or location within the luggage. If an object is determined to contain the characteristics of an explosive, additional slices of the object are collected, in order to determine the mass of the threat.

If the data indicate a threat, the location of the suspicious item is highlighted on both X-ray and CT images. These data are cross-referenced with each other to give the operator an overall image of a suitcase and detailed CT information relating to the contents, and in particular, relating to the potential threat. Using CT at two separate source energies has been proposed to acquire additional information [115]. Since the volume element ('voxel') dimension is known from the geometry of the CT scanning system, the attenuation coefficient becomes a function of only two variables: effective atomic number and density. Separation of these variables is done by taking CT data at two separate source energies, and essentially solving two equations with two unknowns. Thus, the volume elements can be mapped according to effective atomic number and density, and correlation of these variables can provide both feature and material identification.

An additional option on CT instruments is 3D imaging. There are two modes of 3D reconstruction [113]:

- Surface reconstruction, which shows only the surface of the object.
- Volumetric reconstruction, which shows the surface of the object in relation to its surroundings.

Three-dimensional reconstruction is simply the result of computer manipulation of CT scan data obtained in a routine manner.

3.3.1.8 Various X-ray systems

A dual-view dual-energy system has been proposed. The second view will help resolve ambiguities that arise from objects hidden behind others in the first view. Another proposed system consists of five inputs: high energy transmission, low energy transmission, high energy back-scattering, high energy forward-scattering, and orthogonal view high energy transmission [116]. The software will consists of five modules: image registration, image segmentation, region geometric reasoning, region overlap detection, and information grouping. The system will be capable of overlap detection and obtaining geometric information about each object.

3.3.2 Gamma (γ) rays

Gamma rays are emitted by the excited nuclei of atoms. When the nucleus falls from the excited to the ground state, the additional energy is given off as γ-rays. Since only definite discrete energy levels are possible, the γ-rays emitted have definite energies.

Gamma rays, similarly to X-rays, lose energy in passing through matter in three ways [117]: by the photoelectric effect, by the Compton effect, and by pair production. The photoelectric effect is important for heavy absorbing elements and for low γ-ray energies. The Compton effect is important with light target elements and with γ-rays having energies <3 MeV. Pair production of a positron and an electron is important with heavy elements and γ-rays of high energy. Such a pair production requires a minimal energy of 1.02 MeV. Conversely, when a positron and an electron meet, the two are annihilated, and two γ-ray photons with energies of 0.51 MeV each, are produced.

When an electron comes close to an atomic nucleus, it may have its direction and velocity changed. As a result, the electron loses some energy in the form of electromagnetic radiation. This form of γ-ray electromagnetic radiation is known as bremsstrahlung.

3.3.2.1 γ-Ray system based on pair formation

The concept of this explosives detector is shown schematically in Figure 3.12 [118]. A radio frequency linear accelerator is used to produce an electron beam with an energy of 13.5 MeV. The electrons strike a tantalum or tungsten target and produce bremsstrahlung γ-radiation with a maximum energy equal to the electron beam energy. The γ-ray photons interact with the explosive in the examined bag and activate the nitrogen according to the following reaction.

$$^{14}\text{N}(\gamma, \text{n})^{13}\text{N} \longrightarrow {}^{13}\text{C} + \beta^{+} \qquad (3.4)$$

The stable nitrogen isotope ^{14}N thus becomes the radioactive isotope ^{13}N, which then decays with a 10 min half-life via the positron emission to ^{13}C. The

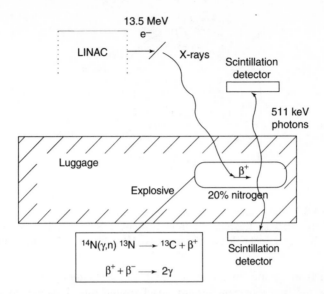

Figure 3.12 Concept of γ-ray system based on pair formation. Reproduced from Clifford, J. R., et al., First Int. Symp. on Explosive Detection Technology, Atlantic City, NJ, 1991, p. 237, with permission.

positron immediately slows down and annihilates, producing two coincident 511 keV photons that are oppositely directed. These photons are detected and counted in coincidence, using standard scintillation detectors.

The high-energy γ-ray photons penetrate easily through most materials and are, therefore, excellent probes for inspecting luggage and cargo. The photons are attenuated significantly only by Compton scattering and pair production. The 10–15 MeV photon attenuation coefficients, μ, range from 0.018 cm²/g for low-Z elements (C, N, O, Al) to 0.030 cm²/g for Cu. Assuming that the luggage examined contains mostly low-Z elements, and that the bulk density is 0.2 g/cm³, the attenuation through 75 cm of luggage will be only about 25%. The system has good resolution because the two oppositely-directed, coincident, 511 keV photons, are counted simultaneously. The explosive lies on the line between the two detectors measuring the event. As many nuclei decay, there are multiple lines that intersect at the explosive's location, as shown in Figure 3.13. Despite the positron range (the distance that a positron travels before annihilation), Compton scattering of the annihilation photons before exiting the luggage, and uncertainty in the location of photon interaction within the detector crystal, the resolution was found to be about one centimeter. This type of imaging is called positron emission tomography (PET).

Since the nitrogen in the explosive has a photoneutron threshold which is substantially lower than that of the most common elements, proper adjustment of the electron beam energy allows activation of the nitrogen in the explosive

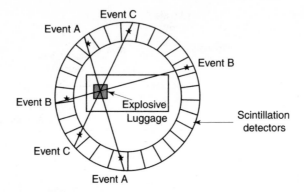

Figure 3.13 Schematic diagram showing the concept of coincidence counting of oppo-sitely-directed photons. Reproduced from Clifford, J. R., et al., First Int. Symp. on Explosive Detection Technology, Atlantic City, NJ, 1991, p. 237, with permission.

without activating the nuclei of most of the surrounding materials. It was found that the best nitrogen-signal-to-background ratio is achieved when the electron beam energy is about 13–14 MeV.

3.3.2.2 γ-Ray nuclear resonance absorption (NRA)

The system is based on detection and imaging of regions of high nitrogen content in the inspected object, by scanning with a high energy gamma ray beam and measuring the transmission profile of the photons [119]. The NRA method uses the 9.17 MeV level in ^{14}N which has a total width of 122 ± 8 eV, so that temperature broadening is negligible. This state is excited by simple proton absorption. Figure 3.14 shows schematically the principle of operation of the detector: 1.75 MeV protons are absorbed by ^{13}C, creating ^{14}N* in the resonant state at 9.17 MeV. The energy of the de-excitation γ-rays is uniquely correlated with the angle of emission with respect to the recoiling ^{14}N*. A resonant flux of 9.17 MeV γ-rays is obtained on the periphery of a cone having a polar angle of 80.7° with respect to the proton beam. Only photons emitted at 80.7° to the direction of the protons will have the correct energy to excite the nitrogen nuclei in the inspected luggage and undergo resonant absorption. These resonant photons are emitted with axial symmetry relative to the proton beam, forming a 0.7° wide cone centered around 80.7°.

The amount of absorption is a measure of the projected nitrogen concen-tration in the luggage.

Production and resonance of the γ-rays can be summarized as follows:

Production: $\qquad ^{13}\text{C} + \text{p} \longrightarrow {}^{14}\text{N}^* + \gamma(9.17\,\text{MeV})$ (3.5)

Resonance: $\gamma(9.17\,\text{MeV}) + {}^{14}\text{N} \longrightarrow {}^{14}\text{N} + \gamma(9.17\,\text{MeV} - \text{attenuated})$ (3.6)

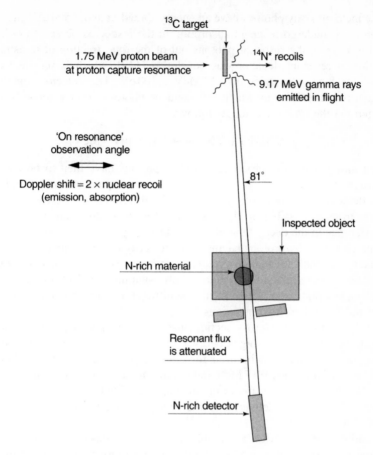

Figure 3.14 Principle of operation of γ-ray nuclear resonance absorption (NRA). Reprinted from Vartsky, D., et al., *Nucl. Instr. and Meth. In Phys. Res. A*, **348**, 688, © 1994, with permission from Elsevier Science.

As a proton source, an accelerator is required, capable of delivering a proton beam with a current of several milliamps, to a ¹³C target. The target must satisfy the following conditions:

- It must contain a ¹³C layer, thick enough to integrate the proton capture resonance yield over all proton energies present in the beam.
- It must also provide nonresonant photons having an energy close to the 9.17 MeV energy of the resonant photons. The flux of these photons must be intense enough to permit determination of the nonresonant component of attenuation, together with the total attenuation measurement.
- It must have good time stability under sustained bombardment of high intensity proton beams.

As the incident γ-ray photons have an energy spread of about 500 eV, only part of them will undergo resonant absorption in the inspected object. In order to select efficiently the resonant photons out of the total fraction of transmitted radiation, a resonant detector was developed [119], which is sensitive only to photons having energies of 9.17 MeV $\pm$ 100 eV. The detector contains a nitrogen-rich medium, in which the incident photons react resonantly with nitrogen via the (γ, p) reaction as follows:

$$\gamma(9.17\,\text{MeV}) + {}^{14}\text{N} \longrightarrow {}^{13}\text{C} + \text{p}(1.5\,\text{MeV}) \tag{3.7}$$

The resulting internally produced 1.5 MeV protons have then to be counted with high discrimination against the numerous Compton electrons produced in the detector by photons of all energies. The differentiation between protons and electrons is possible on the basis of pulse shape discrimination, that is, the difference in the decay time of the produced light in the liquid scintillator, the one produced by protons having a larger decay time. As the proton signal indicates the total attenuation (resonant and nonresonant), the nonresonant attenuation is determined by simultaneously counting the electron signal. The net resonant attenuation, which forms the nitrogen radiogram ('nitrogram'), is calculated from these two quantities.

The large absorption cross section makes the method sensitive to small amounts of nitrogen. The high-energy γ-rays have a great penetrating power so that the technique can be used for quite large containers. The technique produces both a normal 9.17 MeV radiogram and a 'nitrogram', which represents the nitrogen distribution in the luggage examined.

The radiation dose delivered to the object during the inspection is low ($<100\,\mu\text{rad}$).

A variation of the NRA technique has been developed for less common chlorine-rich explosives, such as chlorate and perchlorate mixtures, where the chlorine density is examined instead of the nitrogen density [120]. γ-ray-resonant chlorine (${}^{35}\text{Cl}$) nuclei are emitted, when a proton beam hits a sulfur target, at a 0.5° wide gamma cone at an angle of 82° to the proton beam. The γ-ray photons pass through the inspected material and are resonantly absorbed. The nonabsorbed attenuate beam falls on an arc of bismuth germinate detectors. The nonresonant γ-rays are collected at off-resonant angles by the same detector array and used to obtain total-density data.

3.3.3 Nuclear magnetic resonance (NMR) spectroscopy

In nuclear magnetic resonance (NMR) spectroscopy, the characteristic absorption of energy by certain spinning nuclei (e.g. ${}^{1}\text{H}$, ${}^{13}\text{C}$, ${}^{19}\text{F}$, ${}^{31}\text{P}$) in a strong magnetic field, when irradiated by a second and weaker RF field, perpendicular to it, permits identification of atomic configurations in molecules.

Absorption occurs when these nuclei undergo transitions from one alignment in the applied field to an opposite one.

3.3.3.1 Physical principles [117, 121, 122]

Atomic nuclei having an odd number of neutrons or protons, spin around their axes. As the electrical charges in a nucleus describe an orbit about an axis, the moving charges give rise to a magnetic field lying along the axis of rotation. The resulting magnetic field of each nucleus may be represented by a magnetic moment, μ. As a rotating mass, the nucleus possesses a spin angular momentum directed along the axis of rotation. According to quantum theory, the angular momentum of the nuclei can have only certain values, given by the following equation:

Angular momentum: $\qquad\qquad P = Ih/2\pi$ $\qquad\qquad\qquad$ (3.8)

where: h = Planck's constant
$\qquad I$ = the nuclear spin, has only integral or half-integral values
$\qquad\qquad$ from 0, 1/2, 1, 3/2, 2, etc.

Every isotope has a characteristic spin number. $I = 1/2$ for ^{1}H, ^{13}C, ^{19}F, and ^{15}N. Nuclei with even mass number have zero spin (when both the number of neutrons and protons is even) or spin of an integral value of I. Nuclei with odd mass number have half integral values of I.

When a spinning nucleus is placed in an external magnetic field, as, for example, between the poles of a large magnet, the field exerts a torque upon the nuclear magnet and the nucleus tends to align its magnetic field with the external field. Since only definite orientations of the magnetic field of the nucleus with respect to the external magnetic field are permitted by the quantum theory, the nucleus may begin to precess around the direction of the applied external magnetic field (Figure 3.15). In doing so, the axis of the nucleus tilts and, in addition to spinning about its own axis, carries out a slow

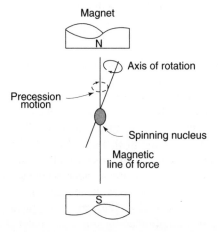

Figure 3.15 The nucleus in a magnetic field.

rotation around the direction of the external field. The stronger the disturbing torque, that is, the stronger the external magnetic field, the more rapid the rate of precession. Hydrogen nuclei placed in a strong, fixed magnetic field of 10 000 G (gauss) will precess 42 million times per second (42 MHz). This rapid precession takes place in every hydrogen nucleus in a sample exposed to the magnetic field. Precession frequencies of most magnetic nuclei are in the range 0.1–40 MHz for fields from 1000 to 10 000 G. The magnetic moment of the nucleus is proportional to the angular momentum and is given by the following equation:

Magnetic moment: $\qquad\qquad \mu = \gamma P = \gamma I h / 2\pi$ $\qquad\qquad$ (3.9)

γ is a proportionality constant called the gyromagnetic ratio, and is different for every isotope, thus permitting separation and identification.

If a nucleus is placed in a magnetic field, there is an interaction between the magnetic moment μ and the magnetic field B. In other words, the energy of the nucleus changes according to the size of the magnetic moment and the angle between them. As this angle cannot have any value, but only a series of discrete values, the interaction energy will also have discrete values, or energy levels. The difference between two energy levels is:

$$\Delta E = \mu H / I \qquad\qquad (3.10)$$

If we apply an electromagnetic field having an energy of ΔE, photons will be absorbed, that is energy will be absorbed by the nuclei. The frequency, ν_0, of the absorbed photons, called resonance frequency, or Larmor frequency, in the magnetic field B will be:

$$\Delta E = h\nu_0 \qquad\qquad (3.11)$$

Combining Eqns 3.9, 3.10, and 3.11 will give the following result:

$$\nu_0 = \gamma H / 2\pi \qquad\qquad (3.12)$$

This is the basic equation of the magnetic resonance method. Table 3.1 shows NMR properties of some selected isotopes

As we can determine the intensity of the magnetic field, we are able, by scanning the frequency of the transmitted RF field (applied perpendicular to the magnetic field B), to reach resonance conditions (Eqn. 3.12) and obtain an absorption signal. The detection of this signal is based upon electromagnetic induction. The sample containing the nuclei under observation is surrounded by a few turns of wire. The resonating nuclear magnetic moments are capable of inducing a small, but detectable RF voltage signal across the terminals of the coil. Figure 3.16 shows a schematic diagram of a typical NMR spectrometer.

A spectrum of signal amplitudes for different nuclei is obtained when the magnetic field over the region of the sample is changed linearly through the

Table 3.1 NMR properties of some selected isotopes

Element	Isotope	Natural abundance (%)	Spin	Gyromagnetic moment, $\gamma/10^7$ $(T^{-1}s^{-1})$	Resonant frequency, ν_0 at 9.395 T (MHz)
Hydrogen	^1H	99.985	1/2	26.75	400.00
Deuterium	^2H	0.015	1	4.11	61.402
Carbon	^{13}C	1.108	1/2	6.73	100.577
Nitrogen	^{14}N	99.63	1	1.93	28.894
Nitrogen	^{15}N	0.37	1/2	2.71	40.531
Oxygen	^{17}O	0.037	5/2	3.63	54.227
Fluorine	^{19}F	100.0	1/2	25.18	376.308

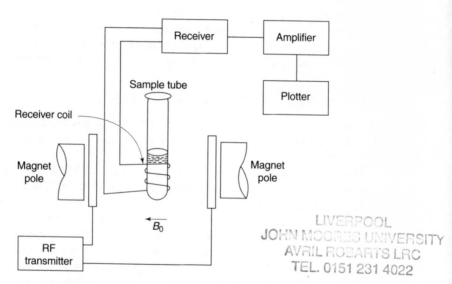

Figure 3.16 Schematic diagram of a typical NMR spectrometer.

resonance value. A spectrum may also be obtained by holding the magnetic field constant and varying the frequency of the oscillating field.

In different molecules, and even within one molecule, the electronic configuration of a nucleus under observation will differ slightly, reflecting differences in the chemical bond. The electron cloud surrounding the nucleus is modified, and the result is an alteration in the precessional frequency. The resulting displacement of the resonance line, when the same nucleus is observed in the same external magnetic field, but in different types of bonds, is called a 'chemical shift'. The resonance spectrum is split into a number of components equal to the number of different chemical groups in which the nucleus is bound. If a proton is not coupled, or interacting, with another proton, a single peak, or singlet, is observed. A proton that is coupled with just one other proton will show a double peak, or doublet; a proton coupled with two other protons, a triplet, etc.

Additional concepts in NMR spectroscopy [123, 124]:

Relaxation: caused by either spin–spin or spin–lattice processes. The rate of relaxation is related to the molecular dynamics of the material.

Spin–lattice relaxation time (T_1): the characteristic time for the nuclear spin system to come to equilibrium with its surroundings following a disturbance. Disturbances may be induced by an RF pulse or by a change in the applied field. It characterizes the rate that the z-component of the magnetization returns to its equilibrium value.

Spin–spin relaxation time (T_2): the characteristic time for the spin system to come to internal equilibrium following a disturbance. It characterizes the rate of decay of the magnetization in the x–y plane.

Coupling constant: also called J-coupling, spin–spin coupling, or scalar coupling. Arises from indirect spin–spin coupling between two nuclei, generally via their bonding electrons. J-coupling is sensitive to bond geometry.

3.3.3.2 Application of NMR to detection of explosives

Hydrogen transient magnetic resonance (HTNMR) has been suggested for detection of explosives concealed in parcels, letters, and airline baggage [125, 126].

The choice of the magnetic field intensity used for NMR is based on trade-offs between the improved sensitivity available at high fields and the greater size, weight, and cost of the required magnet structures, and the increased potential for damage to items being inspected, as the flux density is increased. For detection of explosives, the field intensity is made as low as possible, subject to having enough sensitivity to detect the quantities of concern. The item being inspected is located in a magnetic field of selected intensity, H_0, and tested with an electromagnetic field having a frequency, ν_0, corresponding to the nuclear resonance (Figure 3.17). In the transient mode of operation, as used for explosives detection, the RF electromagnetic field is applied in short pulses of controlled width and amplitude. Detected NMR responses are in the form of transient RF signals emitted by the excited nuclei, following the burst of transmitted energy. The frequency of the emitted NMR signal is that of nuclear resonance in the applied magnetic field, H_0, and the peak amplitude is proportional to the number of nuclei contributing to the response. The amplitude of the transient, free induction decay signal, following a single transmitter pulse, decreases at a rate which, in a homogeneous magnetic field, is dependent upon the spin–spin relaxation time, T_2. This time constant is characteristic of the molecular structure and the state of the sample material. So is the spin–lattice relaxation time, T_1, which sets the time required to detect an NMR response, and the rate at which NMR tests may be repeated without signal degradation. Transient NMR for explosives detection makes use

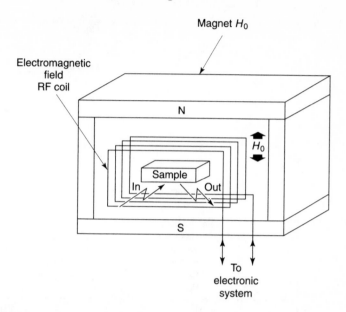

Figure 3.17 Basic concept of explosives detection by NMR. Reproduced from King, J. D., *Advances in Analysis and Detection of Explosives*, Dordrecht, 1993, p. 351, with kind permission from Kluwer Academic Publishers.

of multiple transmitted pulses of appropriate energy and spacing in order to obtain selectivity of the relaxation time constants T_1 and T_2 and the ^1H–^{14}N cross-coupling properties of the explosives.

Figure 3.18 shows the typical range of HTNMR relaxation time constants for most solid, bulk explosives and for common, nonexplosive materials, likely to be found in parcels letters and luggage. At the measured frequency of 3.0 MHz, T_1 and T_2 of explosives were well separated from the other materials.

An explosives detector based on HTNMR will process the total hydrogen signal to isolate a detectable signal component having the long T_1 and short T_2 characteristic of explosives.

The basic sensitivity of HTNMR is a measure of the smallest quantity of the material of interest that may be theoretically detected. This is determined by the signal-to-noise ratio obtained form the material tested by the HTNMR system. For hydrogen nuclei this ratio is given by the following equation:

Signal-to-noise ratio: $\quad V_s/V_n = (4.35 y N_{\text{H}} \cdot 10^{-30}) \cdot (Q v_0{}^3 / BFV_c)^{1/2}$ (3.13)

where: $\quad y$ = instrumentation factor $\cong 1$

$\qquad Q$ = quality factor of sensor coil

$\qquad v_0$ = resonant frequency of nuclei of interest, Hz

$\qquad B$ = detector bandwidth, Hz

$\qquad F$ = noise voltage factor of detector

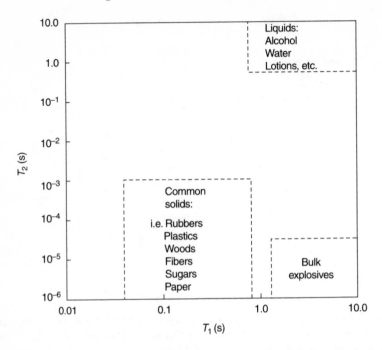

Figure 3.18 Typical 3.0 MHz relaxation time constants for explosives and common nonexplosive materials. Reproduced from King, J. D., *Advances in Analysis and Detection of Explosives*, Dordrecht, 1993, p. 351, with kind permission from Kluwer Academic Publishers.

V_c = volume of sensor coil
N_H = total number of hydrogen nuclei in tested sample

$$N_H = M_s(n/M_{mw})\bullet(6.02 \times 10^{23})$$ (3.14)

where: M_s = weight of sample, grams
M_{mw} = molecular weight of sample
n = number of hydrogen nuclei per molecule of sample

From Eqns 3.13 and 3.14 it is apparent that the signal-to-noise ratio increases directly with the number of nuclei in the HTNMR sensor. It also increases by $\nu_0^{3/2}$ as the resonant frequency is increased by increasing the magnetic field intensity. Thus the minimum quantity that can be detected is affected by the density of the target nuclei in the material of interest and by the sensor parameters.

In an explosives detection system, the luggage passes through a polarizing magnetic field of selected intensity, to remove recorded magnetic media prior to inspection, and then into the inspection magnet.

The HTNMR data required for processing is obtained in a set of multiple sequences. These data allow HTNMR signals from explosives to be recognized in the presence of the typically much larger HTNMR signals from the normal contents of the item examined. The processing algorithms are such that explosives can be detected in many applications with a high probability, while potential false alarms from much greater quantities of nonexplosive materials are greatly suppressed. This is being accomplished by making use of the T_1, T_2 and ^{1}H–^{14}N level crossing characteristics of the HTNMR signals.

3.3.4 Electron spin resonance (ESR) spectroscopy

3.3.4.1 Physical principles [127]

The phenomen of electron spin resonance (ESR), or electron paramagnetic resonance (EPR), is based on the fact that an electron is a charged particle which constantly spins around its axis with a certain angular momentum. Associated with the intrinsic spin is a magnetic moment, the value of which is called the Bohr magneton.

When an external field is applied to the system, the electron will align itself with the direction of this field and precess around this axis. This behavior is analogous to that of a spinning top in the earth's gravitational field. Increasing the applied magnetic field will induce the electron to precess faster. In practice, the magnetic field will split the electrons into two groups. In one group the magnetic moments of the electrons are aligned with the magnetic field, while in the other group the magnetic moments are aligned opposite, or antiparallel, to this external field.

Quantum-mechanically it can be stated that the spin quantum number of an electron is equal to 1/2 and that resolved components of quantum numbers along an axis of quantization must differ by 1. The two possible orientations of these electrons in the applied field correspond to the projections $M_s = \pm 1/2$ along the magnetic field direction. Each orientation is associated with a different energy, the one with the spins antiparallel to the external field ($M_s = -1/2$) being in the lower energy state. These two levels, where the quantum number M_s is either $+1/2$ or $-1/2$, are often referred to as the $+1/2$ or the $-1/2$ states.

If an alternating field is now applied at right angles to the main field, for example, by the use of a high-frequency microwave cavity, an electron can be 'tipped' over when the precession frequency is equal to the incident microwave frequency. ESR can also be described by noting that the quanta of incident microwave radiation may induce transitions between the two states of the unpaired electron. When the energy $h\nu$ of these quanta coincides with the energy level separation $E_{1/2} - E_{-1/2}$ between the two states, resonance absorption occurs. The splitting of energy levels by a magnetic field is referred to as the Zeeman effect. Thus, ESR is the study of direct transitions between electronic Zeeman levels. In practice, the applied electromagnetic radiation is maintained at a certain value, and the magnetic field strength is varied to locate those values, where resonance occurs. The incoming radiation $h\nu$ is absorbed by electrons from both the lower and higher energy levels. The spin population of the ground state, n_1, is larger than that of the excited state, n_2. Therefore, a net absorption of microwave radiation takes place, which is proportional to the population difference $n_1 - n_2$. The population ratio of these two states can be described by the Boltzmann distribution:

$$n_1/n_2 = e^{-h\nu/kT} \tag{3.15}$$

The sensitivity is improved by using a high applied frequency, ν, and a low temperature, T.

In most substances, chemical bonding results in the pairing of electrons to form an ionic bond or sharing of electrons to form an ionic bond. The spins and magnetic moments of paired electrons point in opposite directions and there is no external spin paramagnetism. But in a paramagnetic substance where an unpaired electron is present, resonance occurs at definite values of the applied magnetic field and incident microwave radiation. As with the proton resonance spectrum in NMR, the magnetic behavior of the electron is modified by the magnetic fields in its surroundings. From this deviation from the standard behavior, we can learn about the structure of the studied substances. In an ESR experiment, the sample is placed in a resonant cavity, located between the poles of an electromagnet, at the microwave frequency, ω_0. The magnetic field is then varied until resonance occurs at the value H_0, given by the following equation:

$$\omega_0 = \gamma H_0 = g\beta/\hbar \cdot H_0 \tag{3.16}$$

where: γ = gyromagnetic ratio (ratio of electron's magnetic moment to its angular momentum)

g = 'g-factor', a dimensionless constant (physical property of the electron)

$\hbar$ = Planck's constant h divided by 2π.

β = unit magnetic moment of spinning electron (Bohr magneton)

$$\beta = e\hbar/2mc = 0.92731 \times 10^{-20} \text{ erg/gauss} \tag{3.17}$$

where e is the charge of the electron, m, the mass of the electron, and c, the velocity of light.

Eqn 3.17 constitutes the Larmor condition.

An ESR spectrum is obtained by recording the amount of microwave energy absorbed by the sample as function of the magnetic field.

The resonant absorption has a finite line width, because electrons interact not only with the externally applied magnetic field, but also with the magnetic fields in their environment. By observing line width and line intensity, one can obtain information on electron spin exchange between identical and nonidentical molecules, chemical exchange between the paramagnetic molecule and its environment, and the interactions of neighboring molecules.

3.3.4.2 Consideration of ESR for explosives detection

ESR is limited to a small proportion of materials which have free spins. But when applicable, it has an inherent sensitivity around 15 000 times that of NMR. In practice, this advantage varies widely, depending on relative free-spin concentration, which seldom approaches that of hydrogen, in explosives of interest, and ESR linewidth, which is generally larger than in NMR. However, ESR was found to be exceptionally sensitive to black powder [128]. A representative ESR signal from black powder is shown in Figure 3.19. The sample quantity was 18 mg, and the signal was detected in a commercial spectrometer, operating at X-band, 3000 Oe. The extrapolated signal-to-noise ratio for a 1 g sample would be about 50 000 : 1. When translated to 0.5 kg of explosive at 800 Oe in a scaled-up, luggage-size cavity, one would expect substantial signal-to-noise ratios. Alternately, the field could be greatly reduced for a smaller, but still useful, signal.

Common luggage is not expected to contain materials producing a significant ESR signal, since most organic compounds and plastics are processed in ways which preclude free radicals. On the other hand, black powder, as it does not contain hydrogen, cannot be detected by NMR. It was therefore suggested

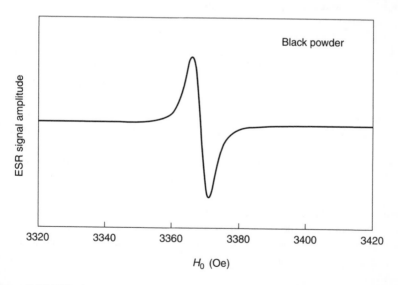

Figure 3.19 ESR signal from black powder. Reproduced from Poindexter, E. H., al., First Int. Symp. on Explosive Detection Technology, Atlantic City, NJ, 1991, p. 493, with permission.

[128], to evaluate the application of an ESR accessory to NMR explosives detectors, for the detection of black powder in nonmetallic luggage.

3.3.5 Nuclear quadrupole resonance (NQR) spectroscopy

3.3.5.1 Physical principles [129, 130]

Nuclei have definite characters of their own which makes them sensitive to the electrical environment in which they may be placed. The study of effects caused by the interaction of nuclei with their surroundings can sometimes give us a detailed picture of the distribution of electrical charge within a molecule. The significant properties of the nuclei are their magnetic moments and their electric quadrupole moments. A nucleus may possess an intrinsic nuclear spin and thus act as a little magnet with magnetic moment, μ. Although a nucleus cannot have an intrinsic dipole moment, it may have a quadrupole moment, if $I > 1/2$. When a nucleus has a quadrupole moment, the distribution of charge in the nucleus must depart from perfect sphericity. We can represent such nuclei as ellipsoids.

Unlike electric dipoles, electric quadrupoles do not interact with spatially uniform electric fields, but only with electric field gradients, which can be understood by regarding the nuclear quadrupole moment as two identical back-to-back electric dipoles.

Many $I > 1/2$ nuclei in low symmetry environments have large quadrupolar interactions and consequently very efficient spin relaxation. The most obvious manifestations of this are the large linewidths generally observed for nuclei such as ^{14}N, ^{17}O, ^{35}Cl, ^{37}Cl.

For example, the ^{14}N resonance of the pyramidal $N(CH_3)_3$ is almost 100 Hz wide, while $^{14}N(CH_3)_4^+$, $^{15}N(CH_3)_3$ and $^{15}N(CH_3)_4^+$ all have nitrogen NMR linewidths less than 1 Hz. These broad lines are produced by the quadrupole splittings and can be observed directly, using nuclear quadrupole resonance (NQR).

In the NMR experiment the splitting between the energy states is induced by a large external magnetic field. As a consequence, gas, liquid or solid can be submitted to NMR spectroscopy. However, for NQR the splitting is due to the interaction of the nuclear quadrupole moment with a molecular gradient in the absence of any external perturbation. Therefore, NQR measurements have to be made in the solid phase at low temperature, since in a liquid or a gas the direction of a molecular bond fluctuates more rapidly with respect to the laboratory referentials than the nucleus can reorient itself with respect to the bond, so that the nuclear quadrupole resonance disappears.

The NQR spectrometer consists of an oscillator, the source for the RF radiation applied to the solid sample in a Dewar, usually at liquid nitrogen temperature, a detector and a data acquisition system. In principle, it is a NMR spectrometer without the magnet. Quantization of the quadrupolar energy levels is effected by the electric gradients available in the molecules, rather than by an externally applied magnetic field.

In summary, NQR spectroscopy is characterized by splittings between the energy states of the order of 10^{-3}–1 cal, with RF frequencies of the corresponding transitions in the range 10–1000 MHz. The NQR frequencies depend on the product of the nuclear quadrupole moment and the electric field gradients [131]. The former is a physical property of the nucleus: all ^{14}N nuclei have the same electric quadrupole moment. The electric field gradients, however, are sensitive to the electronic bonding of the nitrogen, and hence to chemical structure.

3.3.5.2 Detection of explosives by NQR

A NQR explosives detector was constructed and evaluated for the detection of RDX in airline luggage [132, 133]. RDX is one of the most favorable explosives to detect by NQR, as the ^{14}N NQR absorption frequencies from crystalline materials are virtually unique. As NQR signals are weak and are generally below the level determined by the random electrical noise in the RF coil which detects the signal, the signal-to-noise ratio was increased by repeating the experiments as rapidly as possible and adding up the results. The signal-to-noise ratio was thus improved by the square root of the number of repetitions. The maximum rate at which the experiment can be repeated is generally determined by the spin–lattice relaxation time, T_1, which is typically in the range 10–1000 ms for ^{14}N NQR. Hence, the maximum NQR signal would be generated when the experiment is repeated at time intervals greater or equal to T_1. A more efficient alternative is to use methods called steady state free precession (SSFP). The name refers to the steady state condition that occurs when a spin system is irradiated by a continuous train of RF pulses, each separated by an interval τ, during which the nuclear spins precess freely. In pure NQR it was found [134] that the SSFP sequence produces a signal about equal to one-half of the equilibrium magnetization, but at intervals comparable to the spin–spin relaxation time, T_2, of the order of 1–10 ms. Thus, for a given total scan time, a factor of $\frac{1}{2}(T_1/T_2)$ in signal-to-noise ratio can be gained. The performance of the detection system is demonstrated in Figure 3.20, which shows the distributions of the actual intensities of the NQR signals, obtained at 5.2 MHz, in the following four test situations: empty; RDX only; suitcase and contents; suitcase, contents and RDX. For each condition a 6 s scan was repeated for 400 trials (1000 trials for the empty suitcase). The variations seen in signal intensity reflect the random noise contribution to the detector. It can be seen that there are no interfering NQR signals from normal luggage contents and that the suitcase and its contents do not mask the RDX signal. As NQR resonance frequencies are highly specific to chemical structure, signals from other nitrogenous materials did not interfere.

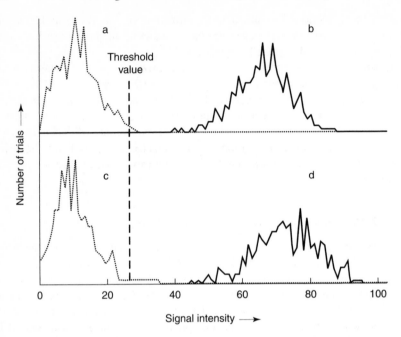

Figure 3.20 Normalized distributions of NQR signal intensities under four test situations: a, empty; b, RDX only; c, suitcase and contents; d, suitcase, contents and RDX. Reproduced from Buess, M. L., et al., *Advances in Analysis and Detection of Explosives*, Dordrecht, 1993, p. 361, with kind permission from Kluwer Academic Publishers.

The system was found to be able to detect subkilogram quantities of RDX-based explosives. An additional feature of an NQR explosive detector is, that since no magnet is required, like in NMR, there is no damage to magnetically recorded material, such as credit cards and computer disks.

A fieldable NQR—renamed quadrupole resonance analysis (QRA)—explosives detector for scanning airline luggage was constructed and tested [135, 136]. Three major factors determine the performance of the system: threshold, signal voltage, and noise voltage. Threshold is a variable influencing sensitivity that can be set arbitrarily to correspond to a certain amount of target material (i.e. the amount of explosive deemed to pose a threat to an aircraft). Signal voltage is the QRA reading on target material, such as explosives. Noise voltage refers to QRA signal readings on nontarget material due to thermal noise and other sources of interference. The objective in doing a QRA scan is to maximize the signal voltage and minimize the noise voltage. However, in some cases, the noise voltage and signal voltage will overlap to a certain degree. A QRA system that has been set up to have a very low threshold may have an excellent probability of detection, but may also have a high false alarm rate. Conversely, a system configured to have a high threshold with a low false alarm rate, may miss detecting the target material with higher frequency.

Figure 3.21 shows the test results on an empty bag and a bag containing a plastic explosive.

One of the tested QRA prototypes used specific transition frequencies of 3.4 MHz for RDX and 0.89 MHz for PETN [137]. The associated relaxation times, T_1 and T_2, for RDX are 11 ms and 0.9 ms, and for PETN 32 s and 0.9 ms, respectively. Because T_1 of PETN is relatively long, only one measurement per scan was performed, whereas many measurements per scan were performed for RDX.

The QR signals from all the measurements during a scan are added, thus increasing the signal-to-noise ratio and improving detection performance. Because of practical limitations on scan time, detection of PETN becomes significantly more difficult, because of the long relaxation time, T_1.

QR searches for the characteristic signals of specific explosive compounds. The QR signal depends on the amount of actual explosive compounds, while

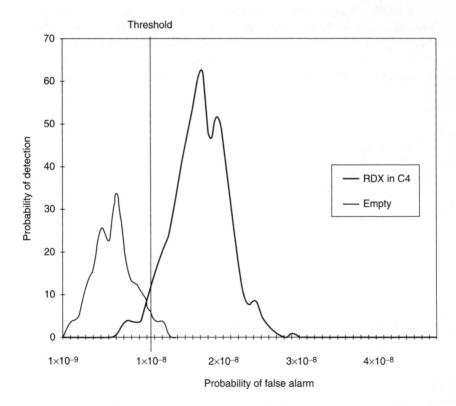

Figure 3.21 Distribution of test results on an empty bag and a bag containing Composition C-4. Reproduced from Rayner, T., et al., Second Explosives Detection Technology Symp., Atlantic City, NJ, 1996, p. 287, with permission.

not responding to other chemicals in the explosive. This makes it more difficult to detect mixtures of various explosive compounds, as opposed to those consisting primarily of only one explosive compound.

A phenomen was observed [137], which caused nonresonance signals, appearing as actual QR signals, or, when strong, obscuring an actual QR signal from an explosive, when present. This phenomen, termed 'acoustic ringing', occurs with certain metals or metal coatings and appears to be due, at least in part, to remnant magnetization of the metal. Items which have been demonstrated to cause this phenomena, include belt buckles, buckles on bags or shoes, jewelry, small scissors or tweezers, chrome-plated sockets, etc. Acoustic ringing was a significant source of false alarms, but has been greatly reduced.

The tested QR explosives scanner could detect RDX and PETN, had a detection coil of 880 liter single copper winding and was automatically tuned to RDX or PETN frequency. Scan time was 4 s for RDX and 3 s for PETN, with the bag tested stationary in the coil. Size of inspected bags was up to 0.8 m (height) × 1.0 m (width) × 1.0 m (length).

An NQR instrument was constructed for remote detection of RDX [138]. The intensity of the NQR signal depends, among other, on the distance between the examined object and the receiving ferrite antenna and on the initial capacity of the exciting generator. The exciting and receiving antennas are located on different sides of the inspected luggage (Figure 3.22). The excitation is

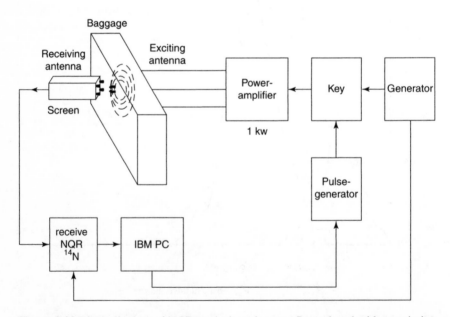

Figure 3.22 Block diagram of NQR explosives detector. Reproduced with permission of author and publisher from Grechishkin, V. S., *Appl. Physics A*, **55**, 505, © 1992 Springer-Verlag Berlin.

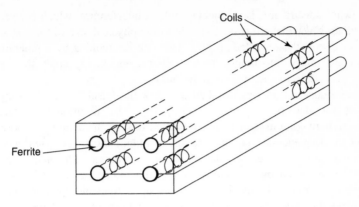

Figure 3.23 Receiving ferrite antenna. Reproduced with permission of author and publisher from Grechishkin, V. S., *Appl. Physics A*, **55**, 505, © 1992 Springer-Verlag Berlin.

produced by a ferrite antenna (Figure 3.23). It was possible to detect RDX and HMX at a distance of 35 cm, using a transmitter power of 1 kW. It was not possible to detect TNT because the NQR frequencies of TNT are considerably lower and the relaxation times longer.

A laboratory study of detection of RDX, using multipulse NQR techniques, was carried out [139]. It was found that the use of special pulse sequences with optimum pulse parameters, could improve the detectability of NQR signals from RDX samples. Results showed that a decrease in detection time could be obtained or alternately, an increase in distance from the sample, in case of remote NQR.

Detection systems including NQR in combination with other technologies have been suggested [140]: in combination with computed tomography (CT), in combination with X-ray, and in combination with a vapor detection system.

3.3.6 Microwave spectroscopy

Microwave spectroscopy provides information on the molecular structure of the sample tested. In backscatter spectroscopy, the test sample is illuminated with microwave radiation of varying wavelength, and the amount of backscattered radiation at each wavelength is measured and the volumetric scattering cross section for the test sample is computed as a function of the wavelength. Absorption and resonance bands, as well as the general scattering level, provide clues as to the chemical nature of the illuminated sample [141]. Microwave spectroscopy differs from atomic spectroscopy in that the absorption and resonance lines are highly broadened. Such broadening, which is due to coupling mechanisms, thermal agitations and dissipative losses at the molecular bonds, reduces the discrimination capability of microwave spectroscopy.

Microwave spectra are also subject to optical interference, which occurs when the radiation wavelength is comparable to the physical size of the test sample and can distort the recorded spectra. This can be avoided by segmenting the microwave image into homogeneous regions, and finally apply the spectroscopic procedure to each region individually.

A laboratory set-up was installed to check the feasibility of the suggested method for detection of explosives [141]. The traditional approach to microwave imaging uses mechanical motion to scan the azimuthal coordinate, pulsed or frequency-swept illumination to obtain the range coordinate, and a receiver array to measure the elevation coordinate. The azimuthal scan is usually achieved by translating or rotating either the target body or the illuminating device. In the laboratory set-up a transmitting horn served to illuminate the target zone, while a receiving horn measured the amplitude of and phase of the back-scattered radiation. The frequency range of the transmitter was 2–18 GHz. The microwave target is illuminated by the transmitting horn at a power of 100 mW. As the target rotates, the detecting system measures and digitizes the back-scattered radiation, which is in turn fed to a data acquisition system, which forms a two-dimensional image of the scattered body.

Ammonium sulfate was used as an explosive simulant. In order to illustrate the wall-penetration capability of the system, a 2 lb box of ammonium sulfate was placed in a soft-sided suitcase and imaged by the transmitter. A monochrome version of the microwave image obtained is shown in Figure 3.24.

A suggested improvement includes the use of higher microwave frequencies, such as 30–100 GHz. It would provide better spatial resolution and a more productive spectral band for differentiating molecular compounds. However, higher-frequency illumination would penetrate suitcase walls less effectively and would thus require higher power illuminators.

3.3.7 Millimeter-wave imaging

Millimeter-waves are electromagnetic waves at frequencies in the range 30–300 GHz, with a wavelengths range of 1–10 mm. As this wavelength is long compared to optical wavelengths, it allows the waves to penetrate many optically opaque materials, such as clothing. Millimeter-waves are typically reflected by the human body and by metals. Dielectric materials, such as plastics, ceramics, and organic materials, will partially reflect the waves and partially transmit them, so that they will appear as partially transparent.

A millimeter-wave imaging system was developed for personnel screening at airport checkpoints [142]. A linear array of 128 antennae (27–33 GHz) is used to scan over a horizontal aperture of 0.75 m, while the linear array is mechanically swept over a vertical aperture of 2 m. The reflected data from the

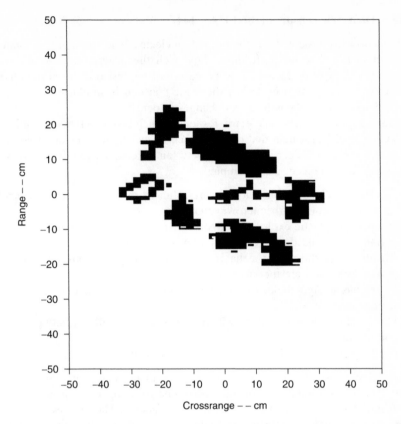

Figure 3.24 Microwave image of a suitcase containing 2 lb of ammonium sulfate. Reproduced from Falconer, D. G., et al., First Int. Symp. on Explosive Detection Technology, Atlantic City, NJ, 1991, p. 486, with permission.

target is collected, using wide-beamwidth antennas, and image-reconstructed. The system has been tested, using concealed metal and plastic weapons, as well as plastic explosives (C-4 and RDX) and simulated explosives, concealed on personnel.

As millimeter-waves do not penetrate the human body, it is necessary to view the subject from several angles in order to carry out a full inspection.

The information that the system presents is simply an image of the object. The system will detect and identify concealed weapons, such as handguns and knives. These weapons may be of metal, plastic, or ceramic. However, it cannot identify the chemical composition of the detected material, and therefore, the operator will not be able to determine specifically that an object is an explosive.

3.3.8 Dielectric sensor method [143, 144]

The dielectric constant, E', is the ratio of an electric field strength in a vacuum to that in tested material. It shows how well the material acts as an insulator. The dissipation factor, D, is a measure of the tested material to convert microwave energy into heat. The dielectric constant is similar to density, so that dielectric changes will reflect changes in density.

The human body has a dielectric response which is uniquely identifiable. Its response is different than that of items such as metals, plastics or explosives. The difference in dielectric responses between different materials can be used to detect threat objects. For example:

- Air has a dielectric constant of nearly 1 and a dissipation factor of nearly zero. It will not absorb microwave energy.
- Plastics and plastic explosives have a low dielectric constant and low dissipation factor, and do not absorb microwave energy.
- Metals can be thought of as having an infinite dielectric constant and dissipation factor and are therefore easily detectable.
- Water has a high dielectric constant (about 80) and medium dissipation factor.
- Human skin is largely water, so its response is about halfway between metal and plastics.
- Clothing has a low dielectric response, similar to air.

The dielectric portal design included a sensor with an antenna/lens combination that establishes a fixed field of microwave energy, at 5.5 GHz, in front of the instrument. Objects which enter that field will alter it. The microwave energy penetrates nonmetallic objects and produces a volumetric reflection coefficient. From the reflection coefficient one can calculate the dielectric constant and the dissipation factor.

The receiver measures the changes in the dielectric characteristics of the field. If an object is placed between the human body and the dielectric sensor, it will give a measurable different dielectric response, which is evaluated by a computer. The dielectric portal consisted of a movable array of 32 dielectric sensors. The person to be screened walks into the portal and stands on designed footprints. And holds onto handle bars at face level. To make a measurement, each sensor is turned on and off, one at a time, and vertical response measurements are taken. The portal then scans the person in a full 360° scan. The scan time is 4 s.

The operation of the portal is similar to a walk-through metal detector, used for the detection of concealed weapons. The screened person declares and removes all items in or under clothing prior to entering the portal. The person then enters the portal, where the scanner completes a 360° scan, and exits. If an item is detected the system will provide information regarding its size and whether it is made of metal or plastic. It cannot define the chemical composition of the detected object.

3.3.9 Raman spectroscopy

3.3.9.1 Physical principles [117, 145]

The Raman effect can be described in terms of transitions between vibrational energy levels or states. When an energetic photon strikes a molecule in its ground state, it may raise the molecule to a higher vibrational state. Since this is not a stable energy state for the molecule, two things can occur now. Most probably the molecule returns to its ground vibrational state and emits a photon with the same energy and frequency as the exciting photon. This is called Rayleigh scattering. However, some of the excited molecules will not return to the ground state, but to some excited vibrational state. Such a molecule emits a photon which has a lower energy than the exciting photon, the energy difference being equal to the difference between the initial and final vibrational states. This is Raman scattering, Stokes type. If a photon is absorbed by a molecule which is in the first excited vibrational state, then the molecule is again raised to some high, nonstable energy state. Most probably, this molecule then returns to the ground state, and in doing so, emits a photon which has a higher energy and a corresponding higher frequency than the exciting photon. The difference in energy between the exciting photon and the emitted photon is equal to the energy difference between the two excited vibrational states of the molecule. This is Raman scattering, anti-Stokes type. Since the probability of a molecule undergoing an anti-Stokes type transition is lower than that of a Stokes type transition. The intensity of anti-Stokes Raman lines is therefore lower than the intensity of Stokes Raman lines.

These phenomena can be represented by the following equations:

$$\Delta E = h\nu_1 - h\nu_2 = h\Delta\nu \qquad (3.18)$$

$$\Delta E = (E_g + E_{vib}) - E_g \qquad (3.19)$$

$$h\Delta\nu = E_{vib} \qquad (3.20)$$

where E_g = energy of ground state
E_{vib} = additional energy of excited vibrational state
ν_1 = frequency of incident radiation
ν_2 = frequency of Raman line
$\Delta\nu$ = Raman shift

The shift in frequency of the lines is proportional to the vibrational energy involved in the transitions. The requirement for an infrared absorption band to appear is that there be a change in the dipole moment of the molecule as it becomes excited. The requirement for the appearance of the Raman effect is that there be a change in polarizability of the molecule.

A molecule subjected to electromagnetic radiation will be polarized because the electric component of the radiation force field will subject the electrons and protons to forces in opposite direction. The value of the induced dipole moment divided by the strength of the field causing the induced dipole is the polarizability of the molecule. This parameter is therefore a measure of the deformability of the electron cloud by the electric field.

3.3.9.2 Evaluation of Raman spectroscopy for detection of explosives

A series of 32 explosives were analyzed by both Fourier transform (FT) and charge-coupled device (CCD) Raman spectrometry in order to determine an appropriate wavelength for the construction of an explosives field detector [146]. The suitability of Raman spectrometry for detecting materials in glass and plastic containers was also investigated.

Two Raman spectrometers were used: a dispersive Raman spectrometer, using either a He:Ne laser operating at 632.8 nm or a diode laser operating at 785 nm; the laser output powers were 20 mW and 10 mW, respectively. The instrument had a silicon, cooled, CCD detector. This system offered a useable Raman shift wavenumber range of $100–4000\,cm^{-1}$ at 632.8 nm and $200–3150\,cm^{-1}$ at 785 nm. The FT Raman spectrometer used a diode-pumped Nd–YAG laser ($\lambda = 1064\,nm$, maximum output 1.6 W) and a room temperature, InGaAs detector. Sample spectra were collected at $4\,cm^{-1}$ resolution over the wavenumber range $\Delta\nu = 200–3600\,cm^{-1}$ by accumulating between 20 and 200 scans.

Results showed that the excitation wavelength of 1064 nm was the best choice for all investigated explosives. Figure 3.25 shows the Raman spectra of C-4, RDX, Semtex, and PETN. However, at this wavelength, for loosely packed samples, sample heating occurs, resulting in thermal emission above $2500\,cm^{-1}$ with a significant baseline offset, presumably because the heat transfer between particles is poor.

In some of the explosives, at wavelengths of 632.8 and 785 nm spectral quality was detrimentally affected by fluorescence.

It was found that both solid and liquid explosives may be analyzed in common glass and plastic containers.

3.3.9.3 In situ detection of explosives particles by Raman spectroscopy

A suggested approach for the detection of plastic explosives containing RDX and/or PETN is to rely on traces of solid residue, microparticles, of RDX and PETN, left inadvertently on various surfaces by a person handling plastic explosives (e.g. the bomb maker), or by fingerprint samples from a person who touched a contaminated object [36].

A Raman microscope–spectrometer system was designed for the detection of traces of plastic explosives in fingerprint samples [147, 148].

The basic idea is to use a high numerical aperture microscope objective as the lens, to focus the laser beam and to collect the Raman scattered

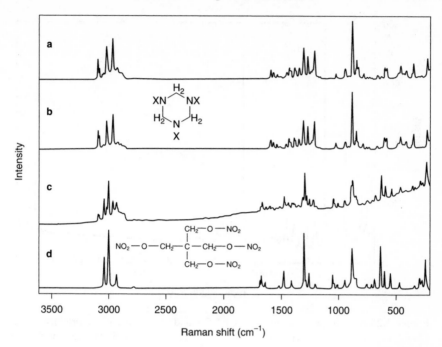

Figure 3.25 Raman spectra of: a, C-4; b, RDX; c, Semtex; d, PETN. Reprinted from Lewis, I. R., et al., *Spectrochim. Acta A*, **51**, 1985, © 1995, with permission from Elsevier Science.

light. The focal spot and/or the size of the sample is of the order of one to a few microns. The optical microscope is an attachment to the Raman microscope–spectrometer system. The system performs two functions: the microprobe takes Raman spectra from micron size samples, to identify the explosive particle, and the microscope produces 2D magnified images of the particle, using only the Raman scattered light.

The instrument covered the spectral range of 100–4000 cm^{-1}. Spectral resolution in the microprobe mode was 1 cm^{-1} and in the microscope mode, 20 cm^{-1}. The Raman imaging field had a 300 μm diameter (with ×10 objective), but could be increased. Spatial resolution of the imaging mode, or the minimal size of the laser focus spot of the spectral mode, was 1 μm. The charge-coupled device (CCD) detector was a two dimensional electronic photographic plate, which had a high quantum efficiency and very low dark current. The instrument, which used a moderate power output (25 mW) laser, was robust and relatively easy to transport and realign.

Typical Raman spectra from micron size RDX and PETN crystals, deposited on a silicon wafer, are shown in Figure 3.26. The measurements were made with a 25 mW HeNe laser, emitting at 632.8 nm. A ×20, numerical aperture (NA) = 0.45, microscope objective was used. The amount of power reaching

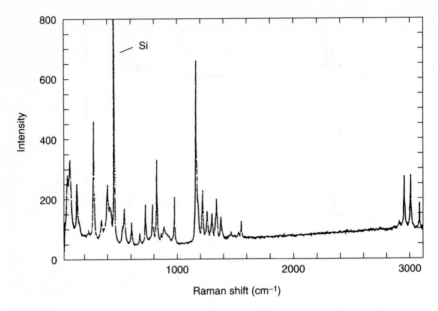

Figure 3.26 Raman spectra of micron size RDX and PETN crystals. Reproduced from Cheng, C., et al., *J. Forensic Sci.*, **40**, 31 (1995). © 1995 ASTM. Reprinted with permission.

the sample was about 5 mW, which corresponds to an energy density of $2 \times 10^9 \, W/m^2$.

Acquisition time for the two spectra was 5 s. Both RDX and PETN appear to have large Raman scattering cross sections, making the Raman spectra easy to acquire. Table 3.2 shows a list of all the Raman bands appearing in the spectra of RDX and PETN.

Raman images of Semtex are usually taken at one or two selected Raman bands which are intense and distinguishable from the background. RDX has a strong band at $885 \, cm^{-1}$ and PETN has a strong band at $874 \, cm^{-1}$, neither of which are polarization dependent. The $20 \, cm^{-1}$ wide imaging filter window can be centered at $880 \, cm^{-1}$ to cover both. Figure 3.27 shows spectra, recorded with different exposure times, from a $5 \, \mu m^3$ particle of Semtex. The spectra confirmed that the particle is RDX.

A fiber optic probe was used for the remote detection and identification of RDX and PETN in Semtex, by Raman spectroscopy. A schematic diagram of the system is shown in Figure 3.28. A 4 m long, $50 \, \mu m$ core diameter, silica optical fiber, supplies the fiber-optic probe head (FOPH) with 633 nm light from a 25 mW HeNe laser. The light is collimated by a singlet lens, and then passed through a line filter. The filter transmits at the frequency of the laser, but stops light having other frequencies, such as light produced by Raman scattering within the fiber. The laser light is reflected by a holographic filter

Table 3.2 List of Raman bands of RDX and PETN. From Cheng, C., et al., *J. Forensic Sci.*, 40, 31, © 1995 ASTM. Reprinted with permission

RDX	(cm^{-1})	PETN	(cm^{-1})
129	S		
152	S	147	S
		194	W
207	S		
226	W	229	S
		260	M
		280	M
304	VW		
		321	W
345	VS,P		
413	M		
463	S	458	M
487	M		
		539	MS
589	W	589	MS
606	M		
		625	VS
668	M	677	M
		705	M
737	W		
		751	W
787	M		
849	M	840	M
		874	VS
885	S		
922	W		
944	W	941	M
		1005	VW
1031	S, P	1033	VW
		1045	M
		1195	W
1217	VS, P		
1235	M, P		
		1253	M
1272	S, P	1279	M
		1294	VS
1312	S, P		
1350	M, P		
1390	M, P		
		1405	W

(continued overleaf)

Table 3.2 *(Continued)*

RDX	(cm^{-1})	PETN	(cm^{-1})
1426	M, P		
		1471	M
1509	VW	1512	VW
1573	VW		
1595	W, P		
		1631	VW
		1662	M
		1675	M
2909	W	2917	W
2953	M		
		2990	M
3006	M		
		3024	M
3079	M		

S: strong; M: moderate; W: weak; V: very; P: polarized peak.

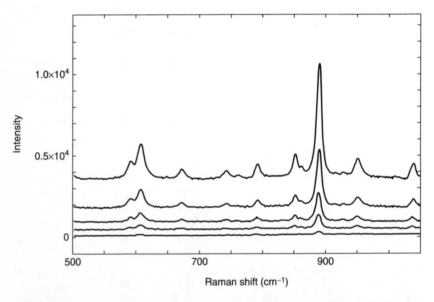

Figure 3.27 Raman spectra from a 5 μm³ particle of Semtex, recorded with 1 s, 5 s, 10 s, and 40 s exposure times (from bottom to top, respectively). Reproduced from Cheng, C., et al., *J. Forensic Sci.*, **40**, 31 (1995). © 1995 ASTM. Reprinted with permission.

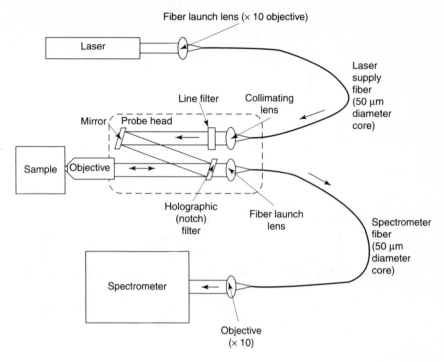

Figure 3.28 Schematic diagram of Raman spectroscopy remote detection fiber optic system. Reproduced from Hayward, I. P., et al., *J. Forensic Sci.*, **40**, 883 (1995). © 1995 ASTM. Reprinted with permission.

and then focused onto the sample by a microscope objective lens. The same objective collects the backscattered light and directs it towards the holographic filter. The filter transmits the Raman shifted light, while preventing the elastically scattered Rayleigh light from reaching the fiber that connects the FOPH to the spectrometer. This reduces the intensity of the reflected and Rayleigh scattered light by 10^5, and prevents the light from generating spurious signals, by exciting Raman scattering within the fiber. The polychromatic Raman scattered light is coupled into the 50 μm diameter spectrometer fiber by an achromatic lens. The end of the 4 m long fiber from the FOPH is connected to the Raman spectrometer. The sample was mounted in front of the FOPH objective, and moved laterally while the spectrometer monitored the intensity of Raman scattered light in the range 400–1300 cm^{-1} (using 0.1 s exposures of the multichannel CCD detector). The movement was stopped when an intense peak at 880 cm^{-1}, from RDX or PETN, was detected. A spectrum was then recorded over the range 200–3200 cm^{-1}. Figure 3.29 shows the spectrum of a Semtex fingerprint as compared to the spectrum of a standard of PETN. The additional background peaks in the spectrum of the fingerprint are probably due to trace amounts of luminescent finger grease or/and resin from the Semtex.

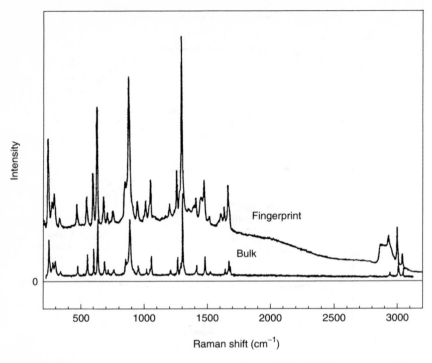

Intensity

0

500   1000   1500   2000   2500   3000

Fingerprint

Bulk

Raman shift (cm^{-1})

Figure 3.29 Raman spectra of a Semtex fingerprint and bulk PETN. Reproduced from Hayward, I. P., et al., *J. Forensic Sci.*, **40**, 883 (1995). © 1995 ASTM. Reprinted with permission.

3.4 NEUTRON TECHNIQUES

Active neutron interrogation techniques form another approach to the detection of explosives [103, 149, 150]. These techniques are based on the interactions of neutrons with individual atoms in the inspected object, in order to determine the presence of specific elements characteristic to explosive materials (see paragraph 3.1). They provide highly penetrating probes that generate distinguishable and detectable reaction products from the different elements constituting the various substances present in the inspected object. The key components of such a technique are a source of neutrons, means to tailor the radiation (e.g. slow down the neutrons or collimate them), the tested object (e.g. a suitcase), a detection device (e.g. scintillation counter) and a data acquisition and processing system to make a decision regarding the presence of an explosive. The various elemental nuclei in the tested object react in different ways (e.g. they have different interaction cross sections) with the incident beam of neutrons. Generally, they emit detectable characteristic radiation, such as high energy γ-rays. The intensity, energy and spatial distribution of the detected radiation, their relation to the incident beam of neutrons, and

Table 3.3 Neutron-based explosive detection techniques

No.	Technique	Abbreviation	Probing radiation	Nuclear reaction	Detected radiation	Measures	Paragraph in book
1	Thermal neutron analysis	TNA	Thermalized neutrons	$^{14}N(n_{th},\gamma)N^{15}$	γ-rays	N	3.4.1
2	Fast neutron analysis	FNA	Fast neutrons	$^{14}N(n,n'\gamma)^{14}N$ $^{16}O(n,n'\gamma)^{16}O$ $^{12}C(n,n'\gamma)^{12}C$	γ-rays	C, N, O	3.4.2
3	Pulsed fast neutron analysis	PFNA	ns pulses of fast neutrons	$^{14}N(n,n'\gamma)^{14}N$ $^{16}O(n,n'\gamma)^{16}O$ $^{12}C(n,n'\gamma)^{12}C$	γ-rays	C, N, O	3.4.4
4	Pulsed fast-thermal neutron analysis	PFTNA	Fast and thermal neutrons	$^{16}O(n,n'\gamma)^{16}O$ $^{12}C(n,n'\gamma)^{12}C$ $^{14}N(n_{th},\gamma)^{15}N$	γ-rays	C, N, O	3.4.3
5	Fast neutron transmission spectroscopy	FNTS	ns pulses or continuous fast neutrons	N(n,n)N C(n,n)C O(n,n)O	Neutrons	C, N, O	3.4.5
6	Gamma-to-fast neutron ratio analysis	GFNA	Fast neutrons	N(n,n'γ)N C(n,n'γ)C O(n,n'γ)O	γ-rays and neutrons	Wide range of elements	3.4.6
7	Neutron elastic scatter	NES	Fast neutrons	N(n,n)N C(n,n)C O(n,n)O	Scattered neutrons	C, N, O	3.4.7
8	Associated particle imaging	API	14 MeV neutrons and α particles	N(n,n'γ)N	Coincident γ-rays	All elements except H and He	3.4.8

any additional available information concerning the tested object, are used in the decision process. A list of neutron-based detection techniques is given in Table 3.3.

3.4.1 Thermal neutron activation (TNA) [151–153]

In TNA, the item to be tested is placed in a thermal-neutron flux. Thermal neutrons are generated by slowing down fast neutrons produced by the spontaneous fission of californium-252, emitting neutrons with an average energy of 2.3 MeV, or by an electronic neutron generator with a neutron flux of less then 109 neutrons per second. The neutrons are thermalized in a high density polyethylene cavity which is capable of handling large suitcases. The thermal neutrons are captured and the resulting compound nucleus is formed with excess energy. If this energy is released in a short time, of the order of 10^{-12} s, then the emission is called prompt; these are the reactions of interest here. A nucleus which has absorbed a neutron will emit γ-rays of characteristic energy at a known rate. Thermal neutron activation of nitrogen can be described by the following reaction:

$$^{14}N + n \text{ (thermal)} \longrightarrow {}^{15}N^* \tag{3.21}$$

$$^{15}N^* \longrightarrow {}^{15}N + \gamma(10.8\,\text{MeV}) \tag{3.22}$$

The resulting γ-ray energy is the greatest γ-ray energy produced from a nuclear reaction. The cross section for this reaction is about 66 mb [149]. The γ-rays are detected by an array of 96 NaI scintillation detectors, coupled to photomultiplier tubes. By knowing the intensity and spatial distribution of the detected γ-rays, the flux of neutrons bombarding the inspected item and the reaction rate for the element of interest, the amount of that particular element in the item tested can be determined.

The cross section for the analogous reactions with carbon and oxygen are much lower than the one for nitrogen, thus making it difficult to detect carbon and oxygen.

The most important physical quantity determined by the TNA technique is the γ-ray spectrum produced by the interaction of the neutron radiation with the luggage tested. An example of such a spectrum is shown in Figure 3.30 [151], which shows the γ-ray spectrum produced by 1 kg of C-4 simulant in a typical full suitcase, with the logarithmic scale for the ordinate covering five orders of magnitude. This spectrum demonstrates the basic difficulty of the TNA method, namely that the number of counts in the nitrogen part of the spectrum is only a small fraction (about 10^{-4}) of the total count rate at the detector, which might lead to artifacts in the reconstruction process. Higher γ-ray count rates could be obtained by longer measurement times or stronger neutron sources, but due to operational constraints, such as speed of luggage processing and concern about damaging film or magnetic tape, it cannot be used.

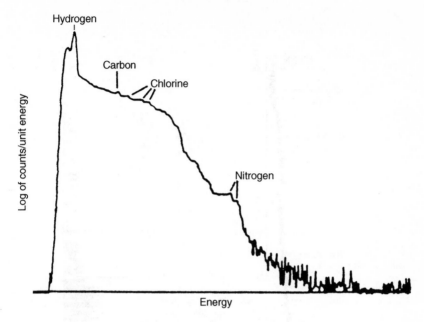

Figure 3.30 γ-Ray spectrum produced by the interaction of neutrons with 1 kg of C-4 simulant. Reproduced from Gozani, T., et al., *J. Energetic Mater.*, **4**, 377 (1986). By permission of Dowden, Brodman and Devine, Inc.

The detection of nitrogen, even in excess of a certain minimal amounts, is not enough to indicate the presence of explosives, because a large number of common materials, such as wool, leather, and nylon, carried in luggage, contain nitrogen. The main difference is that the common materials have a low physical density and are spread over a large volume, whereas explosives are very dense and are very compact in three dimensions. In order to distinguish between common nitrogen-containing materials and explosives, the TNA method creates a three-dimensional image of the nitrogen inside the inspected object, by using algorithms. High concentration of a nitrogen-containing material in a small volume provides a strong indication for the presence of an explosive.

Two TNA systems were tested and compared [154]: one system used a ^{252}Cf radioisotope as a source of neutrons whose energies varied from zero to 10 MeV. The second system used an accelerator to bombard deuterium (absorbed in a scandium target) causing nuclear fusion. This process produced neutrons of a nearly constant energy of 2.5–3 MeV, with a peak neutron intensity of 5×10^8 neutrons/s. Both systems were found to perform equally well in terms of detection/false alarm probability. The advantages of the accelerator were that radiation was produced only when the accelerator was turned on. Also, not having the higher energy neutrons, it reduced the radiation shielding

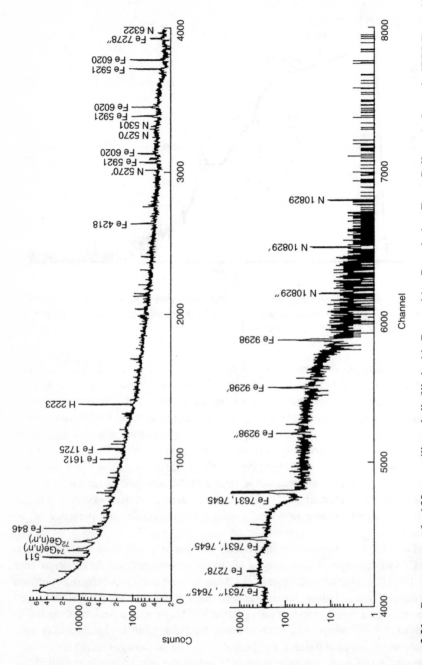

Figure 3.31 γ-Ray spectrum of a 155 mm artillery shell, filled with Composition B explosive. From Caffrey, A. J., et al., *IEEE Trans. Nucl. Sci.*, **39**, 1422 (1992). © 1992 IEEE. Reprinted with permission.

needed for the system. However, the accelerator was a complex and costly device requiring much more servicing and maintenance than the ^{252}Cf source.

The results showed a probability of detection of 90–96% in luggage and of 90–95% in cargo, with a probability of false alarms of 3–8% and 1–4%, respectively. The addition of an X-ray machine with image correlation with the TNA reduced the false alarm rate by 50%.

TNA was used to identify explosives in munitions [155]. The neutron source employed was a two microgram ^{252}Cf source, producing about 106 neutrons/s. A $6 \times 6 \times 3$ inch polyethylene block moderated the neutrons. Neutron-induced γ-rays were counted by a high-purity germanium detector. The method was applied to 155 mm artillery projectiles, filled with composition B explosive (containing RDX and TNT), and 8 inch projectiles, filled with TNT. The measurements detected nitrogen and hydrogen capture γ-rays, as expected from explosives. Figure 3.31 shows the complete γ-ray spectrum of a 155 mm artillery shell, filled with Composition B explosive. The presence of very strong iron capture γ-rays, originating from the steel shell casing, could be used as a built-in energy calibration.

3.4.2 Fast neutron activation (FNA)

While TNA is not applicable to the detection of carbon and oxygen, these elements can be detected by characteristic γ-rays emitted as a result of fast neutron inelastic collisions with isotopic nuclei. These neutrons must have energies above the excitable nuclear levels of the interacting nuclei for γ-rays to be stimulated. Fast neutron interrogation is carried out with neutron energies roughly above 5 MeV and 7 MeV for carbon and for oxygen, respectively. The cross sections for 14 MeV neutrons for C, N, and O are 230, 314, and 474 mb respectively [156]. In order to be used in the elemental detection process, one or more intense high energy γ-rays should be produced. Carbon and oxygen, and to a lesser extent, nitrogen, produce relatively intense γ-rays as the result of nuclear excitation by the (n,n'γ) reaction, as shown in Figure 3.32.

The object inspected is irradiated by a collimated beam of neutrons. The stimulated γ-rays are detected by an array of detectors surrounding the object and shielded from the direct exposure of source neutrons. The neutron interactions in the shielding and the detector materials contribute to the high background inherent to the FNA technique.

Figure 3.33 shows the FNA spectra for a passenger suitcase, where the net signal and background are compared. The imaging in FNA is carried out, as in TNA, from the readings of all the detectors. From the information about the different solid angles each detector subtends at each volume element (voxel) of the tested item and the neutron flux, the amount of specific elements per voxel can be calculated. However, this technique is limited to relatively small objects, like suitcases and parcels, because in large objects there is a lack of geometrical definition, due to the large distance between inner voxels and

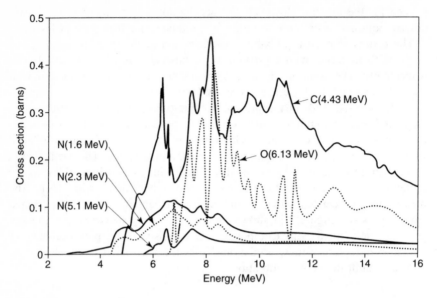

Figure 3.32 Neutron inelastic scattering cross sections for excited states of carbon, oxygen, and nitrogen. Reprinted from Gozani, T., *Nucl. Instr. and Meth. in Phys. Res. A*, **353**, 635, © 1994, with permission from Elsevier Science.

the detectors. Although better collimation of the neutron beam and the γ-ray detectors will improve imaging capabilities, it suffers from a significant loss in sensitivity.

3.4.3 Pulsed fast-thermal neutron analysis

A pulsed fast-thermal neutron system has been developed, which is based on the identification and quantification of O, N, and C in interrogated materials [157]. The three elements C, N, and O are identified through the characteristic γ-rays emitted from the interaction of neutrons with the corresponding nuclei. C and O are identified through the $(n,n'\gamma)$ reaction, while N is identified through the (n_{th},γ) reaction. The combination of fast and thermal neutrons is provided from a pulsed neutron generator, with an output as shown in Figure 3.34. A sealed tube neutron generator with a yield of 10^3–10^4 14.1 MeV neutrons/pulse was used. Using a bismuth germinate (BGO) detector, the carbon and oxygen content of a sample can be identified from the 4.44 and 6.13 MeV γ-rays, respectively. The nitrogen content is identified through the (n_{th},γ) reaction, from the 10.83 MeV γ-rays, which is the highest energy γ-ray produced from any naturally occurring isotope. The disadvantages are the lower cross section for the (n_{th},γ) reaction and the reduced photopeak efficiency of the detector with increasing γ-ray energy. Another problem is the 10.2 MeV γ-ray from the ^{73}Ge(n,γ) reaction in the detector which could

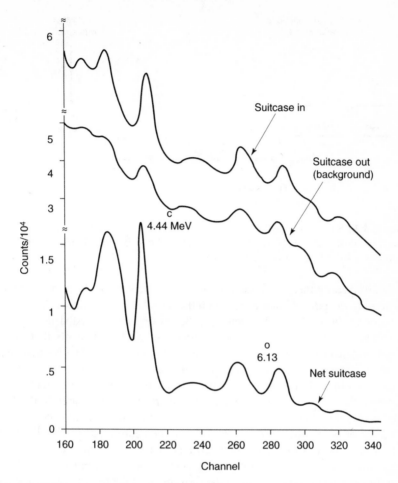

Figure 3.33 FNA spectra for a passenger suitcase. Reprinted from Gozani, T., *Nucl. Instr. and Meth. In Phys. Res. A*, **353**, 635, © 1994, with permission from Elsevier Science.

interfere with the 10.83 MeV γ-ray from the nitrogen. With the right choice of source–detector configuration and appropriate shielding, it was possible to reduce the 10.2 MeV γ-ray so that it would not interfere with γ-rays of interest.

3.4.4 Pulsed fast neutron analysis (PFNA)

Pulsed fast neutron analysis (PFNA) [156, 158, 159] combines two techniques: in-beam γ-ray spectroscopy and neutron time-of-flight (TOF). The known isotopic cross sections for excitation to specific levels, that decay by emitting discrete γ-rays, are used to detect and determine the amount of the specific isotope in the material of interest. The location of the detected material

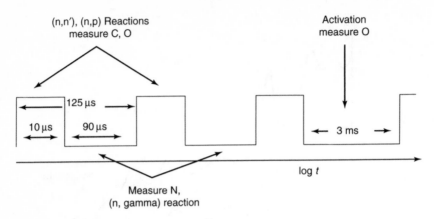

Figure 3.34 Schematic diagram of neutron generator pulsed output. Reproduced from Vourvopoulos, G., et al., *Nucl. Instr. and Meth. in Phys. Res. B*, **79**, 585, © 1993, with permission from Elsevier Science Publishers.

is determined by measuring the monoenergetic neutron TOF from its point of creation to the point of interaction with the tested item. The time of interaction is recorded by an NaI γ-ray detector, placed outside the tested item. PFNA provides directly a three-dimensional image of all the elements that have a significant production of distinguishable and unique γ-rays. The neutrons have to be monoenergetic in order to prevent the faster neutrons from reaching the detector and interfering with the γ-rays stimulated by the slower neutrons. In addition, the neutrons must be bunched in short pulses, compared to the flight time across the object inspected, so that the entire stimulated γ-rays arrive at the detector before the arrival of the front of the neutron pulse. The flight time across the object is of the order of 20 and 60 ns, for a suitcase and a truck, respectively, for 8 MeV neutrons. The neutron beam is limited, by collimation, in the plane perpendicular to the movement of the tested object, and thus defines a pixel. The segment of interaction defines, together with the pixel, the interrogated volume, or voxel. The γ-ray spectroscopy, which is performed within each voxel, provides information on the elemental composition and identifies the materials present there. The smallest size of the image voxel (defining the best spatial resolution) is determined by the size of the collimator opening (defining the pixel) and the neutron pulse width, combined with the detection timing uncertainties. The narrowest neutron pulses-several hundred picoseconds, (equivalent to ≈ 2 cm voxel thickness)-have been obtained, but at the expense of beam intensity. Voxels as small as $5 \times 5 \times 5$ cm were used in luggage inspection experiments.

The energy of the γ-rays produced are characteristic of the material the beam is passing through. 'Signature' γ-rays for carbon and oxygen are 4.44 MeV and 6.13 MeV, respectively, and for nitrogen they are 1.63, 2.3, and 5.1 MeV. The principle of the 3D multielemental mapping is demonstrated in Figure 3.35

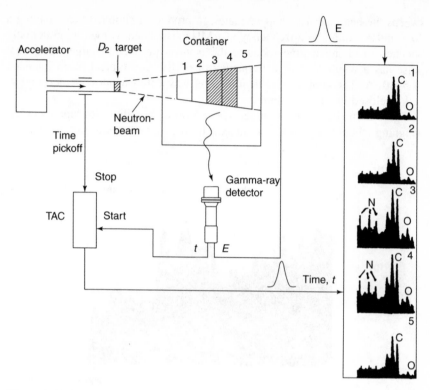

Figure 3.35 Principle of the 3D multielemental mapping in PFN. Reprinted from Sawa, Z. P., *Nucl. Instr. and Meth. in Phys. Res. B*, **79**, 593, © 1993, with permission from Elsevier Science.

[158]. The train of the collimated neutron pulses passes through the sample and produces prompt γ-rays at times T_γ in the indicated voxels, the depths of which are determined by the widths of the time windows ΔT_γ. The γ-ray energy spectra shown on the right are measured in the five indicated voxels. All voxels are filled with fabrics, but voxels 3 and 4 have, in addition a simulant of the C-4 explosive. The elemental densities of C, N and O are derived from the intensities of the characteristic γ-ray lines. The complete 3D elemental distributions are then obtained by directing the neutron beam toward the contiguous pixels on the sample front face. The neutrons are produced by the $D(d,n)^3He$ reaction. The required deuteron energy is in the range 5–6 MeV, depending on the choice of neutron energy and beam energy loss in the neutron production target. This beam was produced by a 3 MV Pelletron tandem van de Graaff accelerator [160]. The source was a source of negative ions by Cesium sputtering, using a titanium deuteride source pellet. The ion source produced a beam of d^- ions having an intensity of up to 120 µA d.c., with a pellet lifetime of 4–8 h. The injection line contained an RF chopper system, which

sweeps the beam across chopper plates, to produce a chopped beam having a 20 ns pulse width and a frequency of 6 MHz. The beam is then longitudinally bunched to obtain a time focus at the neutron production target. The final pulse has a width of 1 ns, a frequency of 6 MHz and typical beam intensity of 5–10 μA. The neutron production target was a thin window deuterium gas cell, enclosed in shield assembly.

Figure 3.36 shows an example, where a portable radio with tape recorder, weighing about 2 kg, was examined. In the upper spectrum, the battery

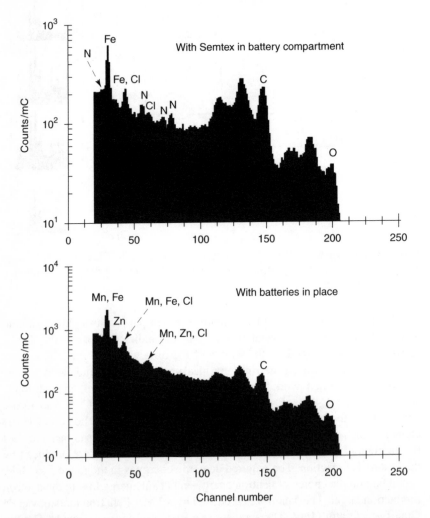

Figure 3.36 γ-Ray energy spectra for a portable radio with Semtex in battery compartment (upper spectrum) and with batteries in place (lower spectrum). Reprinted from Sawa, Z. P., *Nucl. Instr. and Meth. in Phys. Res. B*, **79**, 593, © 1993, with permission from Elsevier Science.

compartment of the radio was partially filled with a simulant of Semtex, while in the lower spectrum, the batteries were reinstalled. Not only was the hidden Semtex simulant detected, but the chemical composition of the batteries was also indicated.

3.4.5 Fast neutron transmission spectroscopy (FNTS)

Fast neutron transmission spectroscopy (FNTS) is based on the fact that many of the light elements of interest, such as carbon, nitrogen and oxygen, have significant cross section features, that allow identification of the elemental constituents in a tested item from the measured neutron transmission ratio [161].

An accelerator is used to produce nanosecond pulsed beams of protons or deuterons that strike a target and produce a pulsed beam of neutrons with a continuum of energies. The material to be tested is placed in the flight path between the target and the detector. Time-of-flight (TOF) techniques are used to measure a transmission ratio as a function of neutron energy. This ratio is unfolded to yield the projected areal densities of the various elements, which are then provided as input to the tomographic reconstruction routines.

A continuous neutron spectrum with energies up to 8.2 MeV is formed by the ^{9}Be(d,n) reaction at $E_d = 4.2$ MeV [162]. The neutrons are collimated in the forward direction into a horizontal fan beam by a 40 cm thick collimator made from high density polyethylene. Backgrounds are minimized by a 1 m diameter target shield of water and lithium carbonate and by a concrete block wall that attenuates background from the target shield. Figure 3.37 shows the layout of the system. An hydraulic lift transports luggage through the fan beam while individual neutron detectors record neutrons transmitted through each pixel. Pixels are 3 cm^2 at the lift location, so that the 16 installed detectors span a slice 48 cm wide. The detectors are 6 cm^2 plastic scintillators 2.5 cm thick,

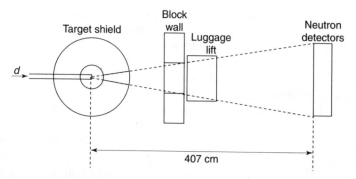

Figure 3.37 Fast neutron transmission spectrometer system. Reproduced from Lefevre, H. W., et al., Conf. Neutrons in Res. & Ind., Crete, 1996, with permission.

coupled to 12 stage photomultiplier tubes, which is connected to a data acquisition system. Transmission spectra are measured and are transferred to the data system, while the location is advanced by one pixel width (3 cm). As the next set of spectra is being measured, the computer calculates neutron attenuations for the previous set, deconvolutes attenuations into projected elemental number densities, and determines the explosive likelihood for each pixel. Additional information includes, for each pixel, an average atomic number, deduced from γ-ray attenuations.

Tests showed that with a time-averaged deuteron beam current of 1 µA, a suitcase 60 cm long (20 pixels) can be automatically imaged in 1600 s. It is believed [162] that this time can be reduced to 8 s by the following improvements to the system:

- Increase of the average beam current by a factor of twenty.
- Reduction of the electronics dead-time by 20-fold or more.
- Increase of the number of detectors to ten rows of 16 detectors.

Various forms of three-dimensional image reconstruction have been proposed to determine absolute densities, in order to single out explosives from the large number of items found in luggage. The suggested algorithms are based on forms of tomography. Classical tomographic techniques need many views for accurate image reconstruction and considerable computation in order to produce an image. These requirements take time and are likely to be very difficult to perform in real time. Consequently, laminographic image reconstruction has been suggested for use with resonant neutron attenuation measurements for luggage inspection [163].

Laminography uses simple image reconstruction, which can be performed in real time, and can work with any number of views. As the number of angles through which an image is viewed increases, the effect of the laminographic focusing also increases and the image quality consequently improves. Simulations were performed, combining laminographic image reconstruction with resonant neutron attenuation. Simulations were made for a variety of suitcases. The contents of the suitcases were chosen to match published distributions of airline luggage. Image reconstruction and discriminant analysis using the simulations were carried out for each of the suitcases for both a variety of explosive types and locations. The simulated suitcase data included statistical fluctuations appropriate to a real neutron radiographic system and a baggage flow rate of 450 suitcases per hour. In order to achieve this throughput and the sensitivity appropriate for explosive detection, a 100 µA deuterium beam bombarding a thick Be target was used in the simulation. This resulted in count-rates in the detectors ($2 \times 2 \times 2.3$ cm plastic scintillators) as high as 1 MHz. The detection efficiency was calculated by determining the number of recoil protons and then only counting those which had a recoil energy greater than 0.67 MeV. The system with these scintillators gave about 1×1 cm transverse resolution in the suitcase. An example of an image of a suitcase with 400 g of TNT is shown in

Figure 3.38. As can be seen, the position of the TNT is well determined in all three dimensions by the laminographic image reconstruction. The displayed three dimensional carbon, nitrogen, oxygen (CNO) feature is a threshold: only voxels with CNO values greater than 6.07, 9.15 and 13.76 are displayed in a, b, and c, respectively.

A ^{252}Cf fast-neutron transmission method for bulk detection of nitrogen-rich explosives has been developed [164]. The technique uses the cross-section

(a)

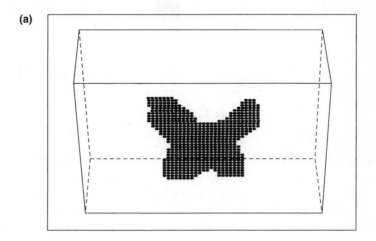

(b)

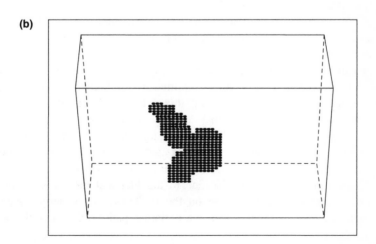

Figure 3.38 3D laminographic density images of 400 g TNT. Voxels with CNO values greater than 6.07, 9.15, and 13.76 are displayed in a, b, and c, respectively. Reprinted with permission of author and publisher from Loveman, R. A., et al., Application of Accelerators in Research and Industry, Denton, TX, 1997. Pp. 895–898. © 1997 American Institute of Physics.

(c)

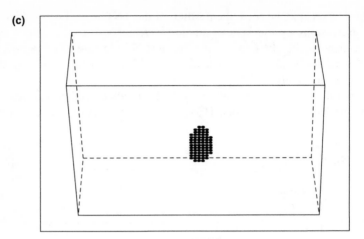

Figure 3.38 *(Continued)*

resonances of nitrogen and oxygen, which are dominant elements in explosives. The cross sections of these elements have characteristic resonances in the neutron energy range from 0.5 to 5.0 MeV, covered by the energy spectrum of the isotopic source ^{252}Cf, which has an average energy of about 2.14 MeV and its energy spectrum spans from 2 to 10 MeV. These resonances are portrayed in the pulse-height distribution of transmitted neutrons, measured with a helium-3 detector.

It is not necessary to determine the exact elemental concentration, as the knowledge of having a high concentration of nitrogen and oxygen is sufficient for a material to be identified as a likely explosive. This high value is reflected in transmission measurements at the resonance energy by the large elemental cross section and further amplified by the higher density of explosives, resulting in a low transmission flux. This amplification process does not exist in a material similar in mass density to explosives, but having a low nitrogen and/or oxygen content. Therefore, a threshold value, determined from experimental measurements can be set at each resonance energy in order to differentiate between explosives and common materials.

A possible problem could arise due to the fact that the scaled transmission flux incorporates information on the thickness of the target material. The thickness of a material within the radiation path could mask the actual explosive.

Two neutron energies, one for nitrogen and one for oxygen, were measured. An energy that corresponds to a resonance in the cross section of either or both elements can therefore act as a good indicator for the presence of explosives. A consistent indication from a pair of energy channels can then be considered as evidence of the presence or absence of an explosive. Ammonium

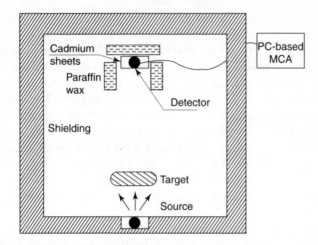

Figure 3.39 Schematic diagram of neutron transmission detector. Reprinted from Gokhale, P. P., et al., *Appl. Rad. Isot.*, **48**, 973, © 1997, with permission from Elsevier Science.

nitrate, representing an explosive, was detected at both nitrogen and oxygen resonance energies, at 1.4 MeV and 3.6 MeV, respectively.

Figure 3.39 shows a schematic diagram of the experimental setup. The source was not collimated in order to cover a large area of the target, thus enabling bulk detection with a few exposures. The isotopic source used in the laboratory experiment contained less than 1 µg of radioactive material, and each transmission measurement was recorded over a period of 15 min. Assuming a 6 s scanning time for a practical system, which is the time needed for inspecting luggage without halting the conveyer belt, the required source would have to contain 150 µg. Since the half-life of ^{252}Cf is 2.65 years, a 300 µg source could be used for 2.65 years without changing the interrogation time.

3.4.6 Ratio of gamma to fast neutron attenuation

A system for the determination of the average atomic number, Z, of heterogeneous mixtures, based on measurement of the ratio of γ-ray to fast neutron attenuation has been suggested [165]. Prompt γ-ray and neutron attenuations were evaluated from neutron time-of-flight (TOF) spectra. Neutrons for these experiments were produced by bombarding a 3 mm thick Be target with a 4.2 MeV deuteron beam from a 5 MV van de Graaff accelerator. The beam was chopped at 1 MHz and klystron bunched. The beam duration at the target was about 1.5 ns, resulting in a time-averaged beam current of 1 µA. Neutrons produced at 0°, with respect to the deuteron beam, were collimated to a fan beam by a 40 cm thick collimator of high-density polyethylene.

Detectors, arranged in a linear array, were positioned about 4 m from the source.

Prompt γ-ray and neutron attenuations were evaluated from the logarithm of the ratio of background-corrected incident to transmitted standardized spectra over an appropriate flight-time (energy) interval. For γ-rays, a single spectrum channel at 3.4 ns/m was used to evaluate the attenuation (which was dominated by Compton scattering), while for neutrons, attenuations were averaged over an energy range of 5.5–8.2 MeV (25–35 ns/m). Samples of various thicknesses of C, Al, Cu, Cd, Sn, and Pb were placed at the exit of the collimator. Results are summarized in Figure 3.40, which shows the ratio of γ-ray to fast neutron attenuations, g_a/n_a as a function of atomic number, Z. Since both attenuations depend linearly on projected number density p, their ratio is independent of sample thickness.

The results of these analyses were incorporated into an explosives-detection algorithm, as the knowledge of Z greatly facilitates the decision about whether the item interrogated is an explosive. In addition, Z has proven valuable in revealing various materials which may shield a concealed explosive from the analysis.

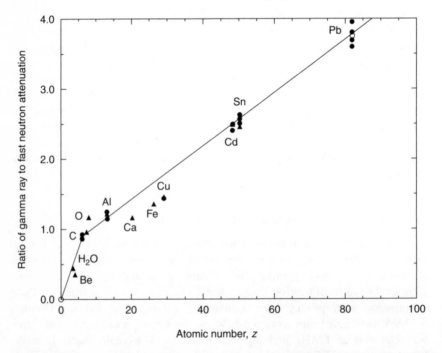

Figure 3.40 Ratio of γ-ray to fast neutron attenuations as a function of atomic number Z. Reprinted from Rasmussen, R. J., et al., *Nucl. Instr. and Meth. in Phys. Res. B*, **124**, 611, © 1997, with permission from Elsevier Science.

3.4.7 Neutron elastic scatter

An explosives detection system, based on neutron elastic scatter (NES) (incorporating neutron resonant elastic scatter (NRES) and neutron elastic back scatter (NEBS)) has been suggested [166]. The detection and location of concealed explosives by NES is based on the measurements of the concentrations of carbon, nitrogen, and oxygen and the calculation of the carbon–nitrogen and oxygen–nitrogen ratios. As can be seen in Table 3.4, the C/N and O/N ratios of the main explosives fall into a fairly narrow range, while their densities are in the range 1.5–2.0 and the concentration of nitrogen is high. Most common materials which contain nitrogen have a lower concentration of nitrogen, and the ratios are different.

Fast neutrons will interact strongly, primarily by elastic collision, with the light nuclei of the target. In elastic collisions there is no change in the structure of the nucleus hit by the neutron. However, neutrons scattered in a certain direction have a new and unique reduced energy, determined by the mass of the nucleus hit. By measuring the change in energy between the incident and scattered neutrons, the nucleus struck can be identified, as all three nuclei of interest produce different back scatter velocities. From the

Table 3.4 Elemental ratios, concentrations and densities of explosives and other materials Reproduced from Gomberg, H. J., and Kushner, B. G., *First Int. Symp. on Explosive Detection Technology*, Atlantic City, NJ, 1991, with permission

	Hydrogen/ nitrogen ratio	Carbon/ nitrogen ratio	Oxygen/ nitrogen ratio	Nitrogen (weight %)	Density (g/cm³)
Explosives (fuel and oxidizer)					
Nitroglycerine	1.67	1	3	18.5	1.70
TNT	1.67	2.33	2	18.5	1.75
RDX	1	0.5	1	38.0	1.6
PETN	2	1.25	3	17.7	1.5
AN	2	0	1.5	35.0	1.7
Commercial articles					
Wool	4.8	3.3	1.1	12	0.2
Silk	4.5	3.0	1.2	15	0.2
Collagen (leather)	4.8	3.1	1.3	15	1.0
Orlon	3+	3+	0	26.4	0.2–0.4
Nylon	11	6	1	12.4	0.2–0.4
Peanuts	14	10	3	4.4	(1)
Navy Beans	13	10	8	4.1	(1)
Melamine	1	0.5	0	66.0	1.5

intensity of the back scattered signals, the amounts of the chemical elements doing the scattering can be calculated, and from the ratio of their signals, the composition can be estimated. For resonance energies, individual elements can show unique changes in cross section, because they no longer are essentially spherical. Instead, they scatter the neutrons more effectively in one direction (back scatter) than another (side scatter), and the cross section may increase several fold, or it may almost disappear, making the element 'transparent'. For example, when the back scatter spectrum for nitrogen is compared with the averaged cross section as a function of neutron energy, important differences emerge. The back scatter resonances are usually higher, resulting in higher sensitivity, but in some cases, such as at 1.59 MeV, there is no back scatter peak. These back scatter resonances can be found at energies up to 10 MeV, which sets an effective energy limit on the NRES technique. The NEBS technique remains effective up to 20 MeV.

An example of laboratory results is presented in Figure 3.41, which shows the spectrum of back scattered neutrons from simulated sheet explosive

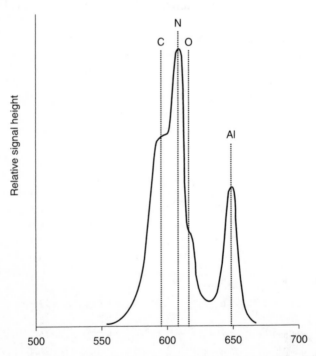

Channel number - relative backscattered neutron energy from time-of-flight

Figure 3.41 Spectrum of back-scattered neutrons from simulated sheet explosive. Reproduced from Gomberg, H. J., et al., First Int. Symp. on Explosive Detection Technology, Atlantic City, NJ, 1991, p. 123, with permission.

(1/4 inch of Melamine + cellulose), covered by a 1/8 inch sheet of aluminum (representing an attache case). Neutron beam energy was 1.78 MeV.

Improvements in the system, in order to reduce false alarms, have been suggested:

- The use of resonance peaks for a critical element like N, followed by off-resonance excitation will cause modulation of the N signature. This will eliminate interfering signals scattered from other elements.
- The normal detector distance for resolution of C, N, O within the signature is 2 m. A second detector at 1 m will display the signature envelope but without resolving the elements. If, however, one component of the signature is from distant scatter, the envelope will change, indicating a false alarm.

The system was found to produce negligible residual radioactivity.

3.4.8 Associated particle imaging (API)

The associated particle imaging (API) technique uses the direction and time correlation between 14.1 MeV neutrons and alpha particles produced in a small accelerator via the deuterium–tritium reaction [151, 167, 168]. Detection of the alpha particle with a position-sensitive detector provides the direction of flight and time of emission of the associated neutron, which interacts with the target nucleus to produce a γ-ray having a characteristic energy of the target material. The direction and subsequent interaction point of the neutron can be determined by the time-of-detection of the γ-ray and is used to locate the neutron–nucleus inelastic interaction and, hence, the target material, while measurement of the γ-ray energy identifies the elemental composition of the target. Typical inelastic cross sections at 14.1 MeV for the target elements found in explosives are several hundred millibarns. Each time-of-flight window corresponds to a well-defined volume element along the direction of the neutron beam for which the γ-ray is measured to determine its elemental composition and density. The system can use the geometric and elemental data recorded with each event to generate real-time, three-dimensional images of identified materials. The API technique will image all elements except hydrogen and helium.

The 14.1 MeV neutrons are produced in a sealed-tube neutron generator (STNG) via the $t(d,n)^4He$ reaction. The STNG also records the alpha particle associated which each neutron, which defines the time of creation and direction of flight for each neutron.

One of the count-rate limitations of the API system is the need to limit random coincidences. If two neutrons are produced in a coincidence window, the system has no way to know which neutron produced the γ-ray. Therefore, the neutron production rate must be kept low enough so that the interactions do not interfere. Thus, the API system is limited to about 2×10^6 neutrons produced traveling in the direction of the target.

Preliminary results showed the imaging capabilities of the system.

3.4.9 Neutron production by accelerators

Various deuteron beam reactions can be employed for producing neutrons. Neutron production yields for various reactions are shown in Figure 3.42 [169]. The abundant $T(d,n)^4He$ reaction is not suitable for TNA application because the 14.7 MeV neutrons produced, are too difficult to shield and produce excessive background.

The $D(d,n)^3He$ reaction has a high yield of neutrons at low energies. The energy distribution of the neutrons emitted from this reaction, at low deuteron beam energy, is almost monoenergetic, resulting in a good neutron thermalization efficiency. The $^9Be(d,n)^{10}B$ reaction can also be used for the production of neutrons. The neutron yield from this reaction is isotropic about the target and its neutron energy distribution is shown in Figure 3.43. It has a high neutron output, but requires higher beam energy. The major advantage of this reaction is that solid beryllium neutron production targets can be used and will last indefinitely. A disadvantage is the production of γ-rays along with the broad neutron spectrum, resulting in a higher background.

The major advantage of the $D(d,n)^3He$ reaction is that neutron yields can be achieved at low voltage or beam energy, but it requires a high beam current. The monoenergetic output results in more efficient thermal neutron production. The main disadvantage is that the target is short lived due to deuterium sputtering erosion which occurs at the high beam currents needed to supply adequate neutron yield.

The Kaman A711 Neutron Generator, using the $D(d,n)^3He$ reaction, produced a beam with a current of 4.0 mA, resulting in a total neutron output of 5×10^8 n/s at an energy of 2.6 MeV. The compact accelerator head consists of an SF_6 insulated pressure vessel containing the glass and metal sealed accelerator tube, consisting of a deuterium ion source, acceleration electrodes and titanium deuteride neutron production target. At the end of the sealed tube, deuterium gas is released by heating a 'getter' material, ionized an focused into a beam, at a potential of 150–200 kV, relative to the target which is at ground potential. The high deuteron current resulted in a limited tube lifetime, due to the high power loading on the target. The useful lifetime of the tubes was generally about 250 h of continuous operation.

The NEC Model 3SH accelerator, using the $^9Be(d,n)^{10}B$ reaction, is contained in a sealed tank filled with SF_6 to insulate the ion source, accelerator column, and pellet-chain high voltage charging system. The maximum beam current was 20 μA at an energy of 1 MeV, with a neutron output of 10^9 n/s. The neutron output was found to be stable and reliable.

The SRL neutron generator was also based on the $^9Be(d,n)^{10}B$ reaction. The positively charged deuterium ions from an RF ion source were accelerated through a compact 600 kV accelerator to a beryllium target. The power

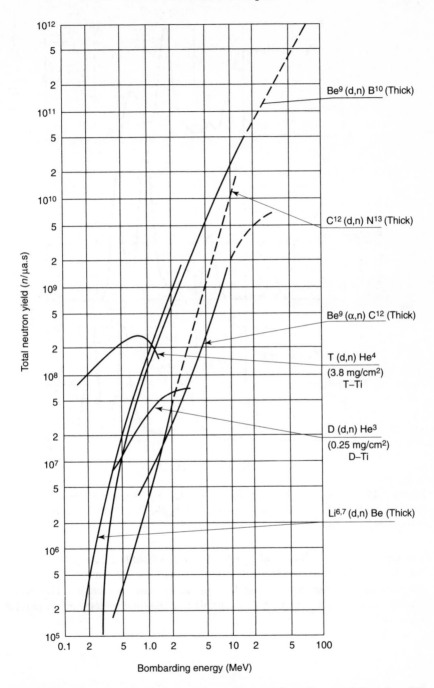

Figure 3.42 Neutron production yields for various reactions. Reprinted from Lee, W., et al., *Nucl. Instr. and Meth. in Phys. Res. B*, **99**, 739, © 1995, with permission from Elsevier Science.

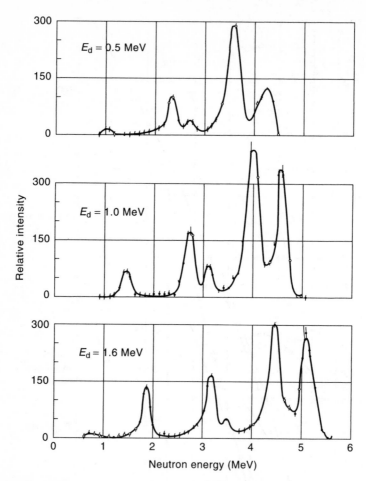

Figure 3.43 Neutron energy distribution from $^9Be(d,n)^{10}B$ reaction. Reprinted from Lee, W., et al., *Nucl. Instr. and Meth. in Phys. Res. B*, **99**, 739, © 1995, with permission from Elsevier Science.

was supplied by a high current cascade voltage multiplier. This power supply was less sensitive to mechanical vibration than an electrostatic generator, which makes it more appropriate for mobile explosive detectors. The beam current was $100\,\mu A$, at $500\,kV$, generating a maximum neutron output of 2.4×10^8 n/s.

4

Systems integration and performance testing

4.1 SYSTEMS INTEGRATION

At present, there is no single detection technique that can by itself, provide a 100% probability of detection combined with a low false alarm rate, that will reduce the threat of terrorism at an acceptable cost to airport operations. A suggested system approach is to combine different technologies having orthogonal detection capabilities [170, 171]. In such a way the strengths of one technique may compensate for the weaknesses of the others, and the vulnerability of one detection device to a potential countermeasure, could be compensated for by another detection device. Clever combinations of independent techniques may achieve detection probabilities and false alarm rates that are more acceptable than those of systems based on one method only. Such a system may even be less expensive than the single method alternative, and could have an acceptable throughput. The performance of such a system will depend largely on the organization of the search strategy, the system architecture, and the operational parameters of the individual devices.

The primary goal of an integrated system is to combine detection devices and procedures, in such a way as to achieve high detection probabilities and throughput rates with an acceptable false alarm rate and cost. The system architecture and search strategy of an explosives detection system (EDS) will vary for each airport situation, due to individual terminal designs, utilization rates, and other local factors.

Three examples of integrated systems are presented. Although all three of them include only bulk detection techniques, integration systems including both bulk and vapor detection techniques cannot be excluded.

First, X-ray enhanced neutron interrogation system (XENIS), which is a combination of thermal neutron activation (TNA) and X-ray radiography [172]. The TNA nitrogen image gives an indication of the threat level within the tested suitcase, but the spatial resolution of the image is coarse, because of limitations on the practical sizes of the gamma detectors. In order to improve the spatial resolution of the nitrogen image, it was correlated with a high-resolution density image. X-Ray radiography is an efficient technique for obtaining a high resolution, 2D density image of an object. Although the X-ray does not provide any information on the elemental composition of the materials in the suitcases, the 2D spatial resolution that it can deliver is better than that obtained with neutron-based systems. The idea of the XENIS is to make use of the most prominent features of each one of the two techniques: the specificity of TNA to nitrogen and the spatial resolution of the X-ray image.

By mathematically correlating the two images, a much more accurate and precise image of nitrogen-containing materials is obtained. The integration of these two pieces of orthogonal information into one explosives detection system is the basis of the XENIS.

The fusion of the TNA and X-ray images is performed by the data reduction module. The first step in the image reprocessing is the conversion of the dimensionality of the TNA image to that of the X-ray image. This is done by projecting the 3D TNA image onto the 2D X-ray image plane. The next steps are image rescaling and image convolution. The size of the reconstructed TNA image is different from that of acquired X-ray image, due to the difference in the data acquisition systems. The sizes of both images have to be matched up before correlation is performed. On the other hand, because of the large contrast in spatial resolution between the TNA and X-ray images, the X-ray image has to go through image convolution, in order to adjust for the resolution mismatch between the TNA and the X-ray images. This is done by filtering the X-ray image, which will lower the image noise level and smooth the X-ray image. After image preprocessing, the TNA and X-ray images are mathematically correlated, using a specially designed algorithm. The most important part in image correlation is to merge the TNA and X-ray images with about the same average sharpness, without losing the important spatial information that the original X-ray high resolution image contains. Another important part is that the intensity of the TNA and X-ray images have to be well balanced according to their significance level. The final image is a correlated image which contains the most important information from both the TNA and X-ray detection systems. In order to obtain an optimal integrated system, additional data has to be included, such as specific airport requirements, and the characteristics of simulants (relative to real explosives) used for the testing of the system.

The detection of explosives (DETEX) system is a multistage multisensor detection system, which combines a dual energy X-ray machine performing 3D reconstruction of the absorption density and a bremsstrahlung source (see paragraph 3.3.2) with a positron tomography system, which can create a nitrogen and oxidation agent density distribution, combined with automatic threat assessment algorithms [173].

The system requirements included a detection threshold of less than one kilogram of explosives, a false alarm rate of less than 0.1% at a detection probability of greater than 99%, and a throughput of at least 600 bags per hour. Explosives to be detected included nitrogen based explosives, such as TNT, RDX, HMX, PETN, Semtex, as well as mixtures based on chlorates and perchlorates.

The DETEX system operates in three stages. In the first stage the bags are activated in an illumination chamber, where the X-ray machine takes three projections from the bag in two energy bands, resulting in an absorption density and an atomic number distribution of the tested bag. An Anger camera (a gamma camera equipped with collimators for computed tomography imaging) is used to produce an activity density map. Mass and activity density are correlated, coherent objects are identified and nitrogen contents assessed. At this stage, unsuspected bags are cleared from the process. In the second stage, suspected bags are transferred into a high precision computed tomograph, which produces a high resolution outline and density map of suspected areas. Bags that can be cleared are taken away from the system. In the third stage, the bags are reactivated at 17 MeV, at the suspect areas only, to avoid excessive activation. At 17 MeV, ^{16}O gets activated and allows assessment of ^{16}O contents and N–O and Cl–O ratios in the suspected area.

Areas which contain enough nitrogen or oxidant carriers like chlorine, and enough oxygen to constitute an explosive charge of sufficient size are considered as bombs and will trigger an alarm to initiate separate treatment.

The third multiple technology system comprises three stages [174]: stage I is a dual-plane, dual energy X-ray in each plane. It determines effective atomic numbers and visual isolation of potential weapon threats. Stage II is an advanced mass–density classification algorithm that uses data generated during the stage I X-ray examination of the target, augments that data with separately processed information and compares the resultant density with known explosive threats. Stage III uses the NQR technology to identify specific explosives. The system has an X-ray anode voltage of 160 kV and contains 256 horizontal and 384 vertical detectors in each plane. The resolution is 0.1 mm.

Simulation models have been developed to assist in developing and assessing various systems alternatives in order to achieve security design goals for screening checked luggage in airports. An example of such a simulation model is shown in Figure 4.1 [175]. The simulation model requires a number of operational parameters describing the environment of the explosives detector, as follows:

- The total number of passengers to be processed.
- The time during which passengers arrive.
- The peak passenger arrival time.
- The average number of checked bags per passenger.
- The probability that a passenger is carrying a bag that contains an explosive.

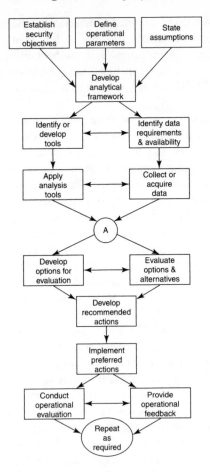

Figure 4.1 Simulation model for explosive detection systems. Reproduced from Smith, M. C., et al., First. Int. Symp. on Explosive Detection Technology, Atlantic City, NJ, 1991, p. 880, with permission.

- The fraction of passengers who will be profiled through interrogation ('manual profiling') or through electronically compiled data ('automatic profiling').
- The first device or procedure to check bags.

In addition to these run parameters, each explosive detector technology and procedure is characterized by a set of parameters, as follows:

- The detection probability of an explosive present in a bag.
- False alarm probability.
- The time required to process a checked bag.

- The next station of an noncleared bag (i.e. an alarm or positive screening result).

The model permits the analyst to select any combination of explosives detector technology and procedures, and configure them in series, parallel, or recycle patterns. The model logic flowchart (Figure 4.2) shows that once a bag is cleared it may be routed to the aircraft. If it fails to clear an inspection, the bag is sent to another device or procedure and, ultimately, to a hand search station for final resolution. If the presence of an explosive is suspected, the bag is sent to a holding area for explosive ordnance disposal.

The simulation model produces a number of results that can be monitored during the simulated period and at the end of each run, as follows:

- The percentage of bags that contain explosives that are detected by the system.
- The percentage of bags that do not contain explosives, that are correctly cleared for boarding.
- The total elapsed time from the arrival of the first passenger until the last one is cleared through the system.
- The single largest number of passengers that were in the system at any one time during the simulated period.

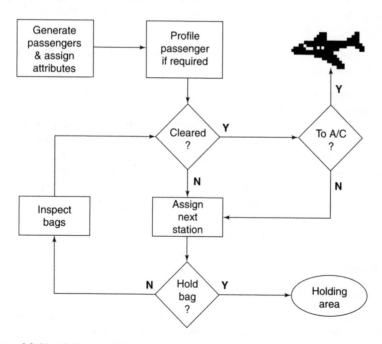

Figure 4.2 Simulation model logic flowchart. Reproduced from Smith, M. C., et al., First. Int. Symp. on Explosive Detection Technology, Atlantic City, NJ, 1991, p. 880, with permission.

Such a simulation can be used to examine various system combinations in an operational environment. It provides a low-cost method for assessing proposed detection techniques, thus allowing several configurations to be tested with minimal change to the overall site.

4.2 PERFORMANCE TESTING AND EVALUATION

Various performance criteria and protocols for testing and evaluation of explosives detection systems have been suggested.

For the purpose of evaluating the stage of development for various instrumental techniques, the following phases were used [171]:

- Concept: description of the concept, including proof-of-principle calculations and drawings of laboratory experimental set-up.
- Demonstration of principle: measurements with laboratory apparatus including detection of pure standards and interferrants.
- Engineering prototype: integration of detection module with other devices and subsystems, testing of key operational parameters and definition of physical size and facility requirements of a fully capable device.
- Deployable device: finalizing of specifications, operational tests, manufacturing and assembly methods, configuration and software. Availability for integration into a qualified explosive detection system. Determination of price.

The detection of small quantities of explosives is complicated by the following considerations [171]:

- Small amounts of explosives have small signatures regardless of the instrumental method used for detection.
- For a given instrument, with lowering of the threshold, the probability of detection increases, but the probability of false alarms also increases.
- A wide variety of items is packed into luggage, presenting a broad spectrum of random background signals to an explosives detection device.
- An explosives detection system would not be acceptable if it substantially slowed down airport operations, or otherwise adversely affected the flow of passengers through the airport.
- Fully automated detection equipment can minimize the human role, but cannot eliminate it completely, since judgment will still be required to clear all alarms.

Performance testing should be carried out in three stages [176]:

1. Laboratory tests, in which fundamental parameters like sensitivity and crude selectivity are being measured, using pure explosive materials.
2. Semirealistic tests, where an attempt is made to detect the explosives in a realistic situation, without trying to simulate operational conditions. Such tests would include pure explosives as well as likely interferrants.

3. Conductance of field tests under airport operational conditions with a variety of bags containing explosives, interferrants and no explosives. In these tests detection rates and false alarm rates are measured.

The following parameters need to be considered when testing an explosives detector [177]:

- Potential locations of explosive detectors.
  Fraction of detection of explosives, $f(d)$, observed in the operational testing of the detector (for different bag populations and different threats).
- Fraction of false alarms, $f(f_a)$, observed for different bag populations.
- Rate of processing of the bags, (R).
- Reliability, maintainability, availability.
- Initial and annual cost.
- Significant operational constraints (environment, manpower, portability, etc.).

$$f(d) = \text{(number of detections)/(total number of possible detections)} \tag{4.1}$$

$$f(f_a) = \text{(number of false alarms)/(total number of possible false alarms)} \tag{4.2}$$

$$R = \text{(time required to process all bags)/(number of bags processed)} \tag{4.3}$$

Other factors to be considered when testing a system include the following:

- Identification of any countermeasure technique to be included in the testing.
- Identification of the set of threat explosives (type, shape and weight).
- Relationship between $f(f_a)$ and $f(d)$, as a function of the threshold setting.
- Simulant validation data.

Additional tests should include separate controls for particle and vapor detectors and for bulk detectors.

4.2.1 Testing of detection systems for bulk explosives

In addition to the previously tests, the following ones are specific to bulk explosives detection systems [176]:

- Characteristic parameter identification (i.e. nitrogen content).
- Effect of contents of bags on measurements.
- Screening effects of various materials.
- Determination of characteristic bags.
- Response to various masses of different explosives.
- Variation in response with target distance or position.
- Effect of position, size and shape of explosives.
- Radiological safety.

4.2.2 Evaluation of particle and vapor detectors

Explosive vapor and particle detectors are designed to detect indirect evidence of the hazard [178]. Unfortunately, neither the amount of vapor emanating from a packaged explosive device nor the amount of surface residue resulting from handling the device, bears a direct relationship to the quantity of explosive within the package. Factors such as the type of wrapping material used and the method of packaging, can override the influence of quantity. Thus, a small amount of plastic explosive hidden in a suitcase by one person, can result in microgram levels of deposit on the exterior of the suitcase, while a large amount of the same explosive, placed by another person inside the suitcase, may leave only nanogram quantities on the exterior. It is therefore impractical to define trace explosive detection standards simply in terms of the quantity of concealed explosive.

Test procedures for vapor and particle explosive detectors have been suggested, which allow equipment performance to be evaluated in terms of trace levels [178]. The ICAO (International Civil Aviation Organisation) 'standard box' (Figure 4.3) and 'standard suitcase' (Figure 4.4) tests were suggested in 1990 as a well-defined protocol, to be used for the comparative assessment of trace vapor detectors in a particular field scenario [179]. These standard containers have since then been developed, but the updated version has not yet been published in the open literature.

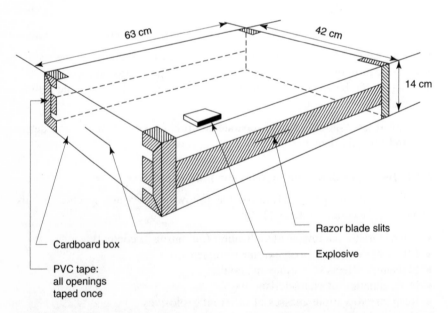

Figure 4.3 ICAO 'Standard Box'. Reproduced from Scharer, J., Third Meeting of ICAO Ad Hoc Group of Experts on Detection of Explosives, Montreal, 1990. Paper AH-DE/3–14, with permission.

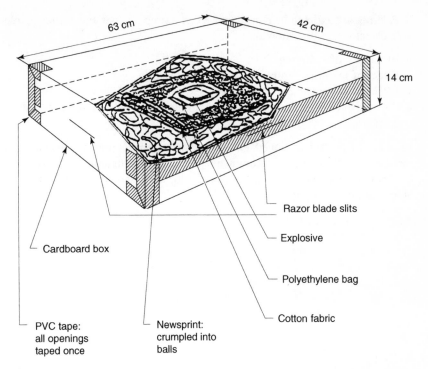

Figure 4.4 ICAO 'Standard Suitcase'. Reproduced from Scharer, J., Third Meeting of ICAO Ad Hoc Group of Experts on Detection of Explosives, Montreal, 1990. Paper AH-DE/3–14, with permission.

Laboratory test procedures for quantifying detection limits of vapor sniffers make use of vapor generators (see paragraph 2.11).

Vapor detectors will not detect plastic explosives, based on RDX and PETN, which have vapor pressures that are several orders of magnitude lower than those of other nitrate ester and nitroaromatic explosives. However, these explosives can be detected as surface contamination with existing detector technology. The problem of providing known, trace amounts of plastic explosives on various material surfaces for testing purposes has been tackled, resulting in a suggested 'calibrated thumbprint' technique [36]. The technique consists of pressing the thumb, the skin of which was previously coated with a thin layer of plastic explosive, in sequence, on a number of stainless steel plates. The odd-numbered plates are rinsed with solvent and analyzed by GC to quantify the amount of explosive on them, while the even-numbered plates are used to test the detector examined. The relative amounts of thumbprint residue in a series can be made to decrease uniformly and slowly, so that the deposits on the even-numbered plates can be interpolated. Aluminum foil and polyethylene sheet have also been found to be suitable materials for providing such calibrated thumbprints.

Additional factors to be considered when testing a vapor and particle explosive detector are the following:

- Efficiency of sampling.
- Effect of impurities in explosive.
- Effect of commonplace chemicals.
- Capability of detection of explosives in electrical devices, in cupboards, in cars.

In order to protect deployed explosives detection equipment against countermeasures, it is highly recommended that the configuration of particular explosives detection devices at a particular location, should not be made available to the general public [171].

5

Tagging of explosives

5.1 INTRODUCTION

Taggants or tags are substances added to explosives at the time of manufacture in order to facilitate the identification and/or detection of the explosives.

'Identification taggants' are designed to be retrieved from the debris of an explosion. They contain a code identifying the manufacturer and the batch of explosives used in the bombing. Hence law-enforcement investigators can discover what kind of explosive compound had been used, and trace the source of that explosive.

'Detection taggants' are designed to facilitate the detection of hidden explosives. The taggant is sensed by a simple detector, even when the explosives are well packaged. The detection taggant is usually a microcapsule emitting small quantities of vapor, which is easily detectable by a simple vapor detector.

5.2 IDENTIFICATION TAGGANTS

The development of tagging has focused on the incorporation of tiny coded particles into explosives during their manufacture [180, 181]. Some of these particles can survive detonation and be recovered and decoded. The identification taggant must carry information necessary to trace the explosive to its left legal holder. Once the manufacturer and the type, date, and batch of the explosive are determined, the explosive can be traced through its commercial channels to the last legal holders [182]. If the explosive has been used in a crime, this trace provides the investigator with several important clues: the investigator knows the type of explosive, and where and approximately when, the explosive passed into illegal hands. When identification taggants are found at the bombing scene, the explosive can be traced in the same way that firearms used in crimes can be traced by the serial number.

In Switzerland, identification tagging of explosives, safety fuses, and detonating cords has become obligatory by law [183]. Two taggants were approved in 1983 in Switzerland:

- *Microtaggant*, invented by 3M Company (St. Paul, MN, USA) and manufactured today by Microtrace (Minneapolis, MN, USA), consists of a multilayered melamine resin lamina, having a maximum diameter of 800 µm (Figure 5.1). The code is represented by nine colored layers, which are recognizable at a magnification of 40 to 100 times. With these nine layers it is possible to create several hundred thousand combinations or codes. A magnetic layer was added to facilitate the collection of taggants in debris with magnets. The amount of this taggant in explosives is 0.025–0.1%.

- *Explotracer*, made by Plast Laboratories (Bulle, FR, Switzerland), is based on colored plastic powder mixed with fluorescent pigments (Figure 5.2). In order to obtain magnetic particles, the basic material is mixed with iron powder. Addition of rare earth elements and other substances (inorganic compounds, such as oxides) permit analytical verification. The taggant can easily be detected with a long wave UV radiation source and can be separated from debris with magnets. The *Explotracer* is a granulate with grains size 200–800 µm, to be mixed with the explosives at an amount of 0.1%. The various components of this substance: the polymer, the fluorescent pigment and the rare-earth elements additional substances, provide the code composition. Ten available forms of each component could result in almost 10 000 combinations. The advantage of this system is the easy detection of the taggants after the explosion. This is due to the relative large quantity of taggants and on their being completely fluorescent. Their magnetic susceptibility is less than that of *Microtaggant*, because the iron particles mix in an irregular way with the granulate. In addition, the time

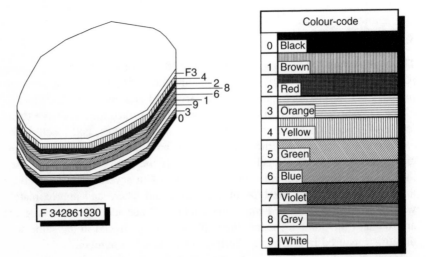

Figure 5.1 3M Microtaggant. Reproduced from Scharer, J., First Int. Symp. on Analysis and Detection of Explosives, Quantico, VA, 1983, p. 463, with permission.

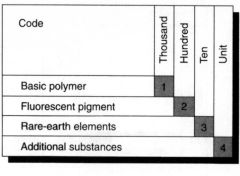

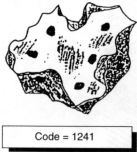

Code = 1241

Figure 5.2 Plast Laboratory Explotracer. Reproduced from Scharer, J., First Int. Symp. on Analysis and Detection of Explosives, Quantico, VA, 1983, p. 463, with permission.

necessary to decode these particles is high, because physical-chemical analyses are needed.

An additional taggant, *ICPM*, made by Synthesia (Pardubice Semtin, Czech Republic), is similar in its concept to the *Explotracer*, with the difference that the particles are flat compared to the granular form of the *Explotracer*. The size of the particles is approximately 0.5 mm, and the concentration of the taggant in the explosive is 0.035 to 0.1%.

The methods used to analyze the different components of the taggant include melting points, thermal analysis and polarizing microscope (for the plastic powder), 300–600 nm UV spectral analysis (for the fluorescent pigments) and X-ray spectrometry (for the rare earth elements and additional substances) Safety fuses are provided with special marking threads. Figure 5.3 shows an example of marking colors for the years 1981–1992. Twelve marking-thread numbers were combined with 12 well-defined colors. By bringing in two threads with the same color, the year is indicated, and with the third colored thread, the month is determined. The marking threads indicating the date of manufacture are found immediately under the plastic coat, because marking threads around the core of the safety fuse cannot be analyzed after it has burned.

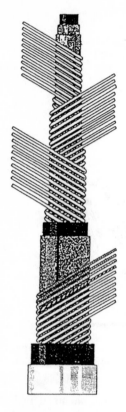

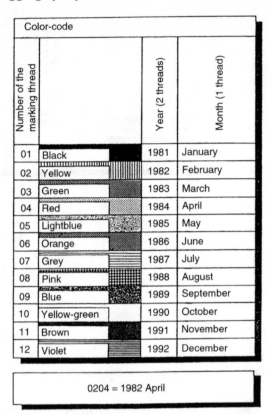

Color-code			
Number of the marking thread		Year (2 threads)	Month (1 thread)
01	Black	1981	January
02	Yellow	1982	February
03	Green	1983	March
04	Red	1984	April
05	Lightblue	1985	May
06	Orange	1986	June
07	Grey	1987	July
08	Pink	1988	August
09	Blue	1989	September
10	Yellow-green	1990	October
11	Brown	1991	November
12	Violet	1992	December

0204 = 1982 April

Figure 5.3 Marking threads and colors for safety fuses. Reproduced from Scharer, J., First Int. Symp. on Analysis and Detection of Explosives, Quantico, VA, 1983, p. 463, with permission.

Explosive cords are marked by additional plastic coats on the cords, instead of the color threads.

5.3 DETECTION TAGGANTS

The required characteristics of a detection taggant have been defined as follows [182]:

- It should not exist in nature. Hence, there would be no required adjustment for ambient background in tests.
- It should continue to release its vapor for at least 5 years, preferably 10 years.
- It should be detectable down to part per trillion (p.p.t.) levels.
- The vapor should not be adsorptive on normal articles in luggage.
- It must not change the properties and performance of the explosive.

Additional requirements were added later on [184]:

- It should be reliably detectable by commercial vapor detectors with minimal false alarms.
- It should not degrade other materials with which they are packaged.
- It should be safely manufactured and stored.

Several of the earlier suggested detection taggants were the following perfluorinated alkanes: perfluoromethylcyclohexane (PMCH), perfluorodimethylcyclohexane (PDCH), perfluorodecalin (PFD), perfluorodimethylcyclobutane (PDCB), and perfluorohexylsulfur pentafluoride (L-4412). These materials have then to be packaged into microcapsules (or microspheres) having a controlled release rate, resulting in a life-span of at least 5 years.

The International Civil Aviation Organization (ICAO) has mandated that volatile taggants be added to plastic explosives during their manufacture in order to facilitate the detection of these explosives [185]. In 1996, the US Congress passed the Anti-Terrorism Bill into law, requiring that plastic explosives be chemically tagged for easy detection. Congress further mandated an extensive study of the feasibility of requiring identification tagging of other explosives and potentially-dangerous products, such as ammonium nitrate fertilizers.

According to the ICAO recommendations, the manufacturer of explosives will select a taggant from a selection of four compounds: ethylene glycol dinitrate (EGDN), o-mononitrotoluene (o-MNT), p-mononitrotoluene (p-MNT), and 2,3-dimethyl-2,3-dinitrobutane (DMNB), at minimum concentrations of 0.2, 0.5, 0.5, and 0.1%, respectively.

The addition of o-MNT, p-MNT, and DMNB was investigated in order to enhance detectability of a German-made, PETN-containing, plastic explosive, Seismoplast [186]. The problem which arose was that none of the suggested tagging compounds was soluble in silicon oil, which is the plasticizer used. This made the mixing of the taggants with the explosive more difficult. Also, the manufacturing process required at some stage, removal of water under a slight vacuum at a temperature of about 60°C, which impaired the quantitation of the amount of taggant added to the explosive.

A program for the tagging of Composition C-4 has been initiated [184, 187]. Composition C-4 was homogeneously tagged with 1 and 0.1 wt % DMNB at a plant laboratory scale, and the parameters of the modified explosive were evaluated. The parameters investigated included homogeneity, detectability, life-time, stability, compatibility, performance, sensitivity, mechanical properties, and toxicity.

The results of both the standard compatibility test and the differential scanning calorimetry (DSC) study on the tagged Composition C-4 were indistinguishable from the untagged control Composition C-4, within experimental error [187]. The compatibility between DMNB and PETN, contained in other plastic explosives, was also tested and they were found to be compatible. The results of the impact sensitivity test, shock sensitivity test, detonation velocity measurements, and mechanical properties tests showed no significant differences between the tagged and the control Composition C-4, within experimental error.

The results of *in vitro* and *in vivo* short-term toxicity tests indicated that DMNB is neither mutagenic nor carcinogenic. The oral LD_{50} for the Swiss-Webster mouse was estimated to be similar to that of RDX for rat. Since RDX constitutes the main component of Composition C-4, the toxicity of DMNB at the 1.0% wt. or lower levels, is acceptable, unless the results of additional long-term tests, which are being conducted, indicate otherwise.

Environmental, as well as manufacture and qualification, testing was also carried out. Additional tests included the ensuring of the uniform distribution of DMNB, using colored material to monitor the dispersal in the mixer and the effect of the taggant on the nonexplosive components in the end items.

Life-time of DMNB in Composition C-4 was studies [188]. The semifinite linear diffusion model was applied to predict the life time of the taggant and was compared with experimental results. The following equation was used:

$$Q_t/Q_0 = (8/\pi^2) \sum_{k=0}^{\infty} (1 - \exp\{-(2k + 1)^2 \pi^2 2Dt/4\partial^2\})/(2k + 1)^2 \qquad (5.1)$$

Where: Q_t = concentration of the taggant at time t

$\quad\quad Q_0$ = initial concentration of the taggant

$\quad\quad D$ = diffusion coefficient of the taggant

$\quad\quad \partial$ = sample thickness.

Figure 5.4 shows the result of fitting Eqn 5.1 to the taggant loss data obtained at 30°C. The model seems to predict a slightly greater taggant loss at the initial stages and a slightly lower taggant loss after longer periods of time.

A method for measuring the emission rate of the taggants EGDN and DMNB from the tagged explosives Semtex 1A and Composition C-4 was developed [189]. The emission rate of taggants from a tagged plastic explosive is important because it will determine the detectability of explosive vapor detectors. Semtex 1A was tagged with 0.2% EGDN. The emission rate of EGDN from Semtex 1A, taken directly from the end of the production line, was found to be 2.52×10^{-5} g/cm·h. After 6 days the emission rate was higher by approximately 50%. This is caused by diffusion of EGDN from the taggants into the explosive, which is a time dependent process. The emission rate depends also on the concentration of EGDN (it is increasing linearly with increasing concentration) and on the composition and physical properties of the plastic explosive: the emission rate of EGDN in Semtex 1A was almost twice that of EGDN in C-4. The emission rate of DMNB from Semtex 1A was found to be much lower than that of EGDN because the vapor pressure of DMNB is lower. The changes in the concentrations of the taggants DMNB and *p*-MNT in the emulsion explosive EMSIT 1, were measured during storage under various conditions, at a function of time [190]. Tagging of emulsion explosives is very important because the detectibility of these explosives with existing

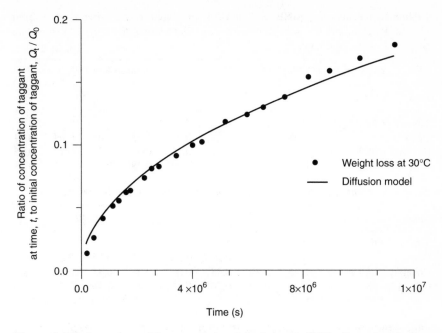

Figure 5.4 Fitting of model to experimental data at 30°C. Reproduced from Reed, R. A., et al., *Advances in Analysis and Detection of Explosives*, Dordrecht, 1993, p. 403, with kind permission from Kluwer Academic Publishers.

methods is very difficult. The guaranteed shelf-life of emulsion explosives is usually only 2 years, which is less than for plastic explosives. The removal of the taggant from the emulsion explosive is not easy, and is usually accompanied by the destruction of the emulsion and consequent loss of explosive properties. Two batches of EMSIT 1 were tagged with DMNB and *p*-MNT, respectively. The tagged explosives were wrapped up in a multilayer plastic foil, forming cylindrical cartridges of 30 mm outer diameter × 320 mm long and 90 mm outer diameter × 360 mm long. The cartridges, weighing 200 g and 3000 g, were placed in open and closed cartons for storage. Simultaneously, laboratory tests were made with the tagged explosives, in open and closed weighing bottles, at temperatures of 20 and 40°C. The taggant concentrations, in both the storage and laboratory tests, were measured at the beginning of the test and during the storage period up to 18 months. The loss of both taggants from commercially packed EMSIT 1 was found to be less than 50% after 18 months of storage, in both open and closed cartons. The emission rates of DMNB and *p*-MNT were in the 10^{-6} and 10^{-5} g/cm·h range, respectively. The changes in emission rates of the two taggants were relatively small during the 18 month storage period.

No differences were found in the concentration of taggants and emission rates in samples taken from the surface and from inside the cartridges,

Table 5.1 Contents of DMNB and *p*-MNT in stored bottles. Reproduced from Mostak, P., et al., *5th Int. Symp. on Analysis and Detection of Explosives*, Washington, DC, 1995, with permission

Content of DMNB in EMSIT 1 stored in glass weighing bottles (%)

Temperature (°C)	20				40			
Time (months)	0	3	6	18	0	3	6	18
Open	0.38	0.23	0.20	0.02	0.38	0.07	0.04	0.04
Covered with PE foil	0.38	0.34	0.31	0.25	0.38	0.27	0.12	0.05
Closed by ground lid	0.38	0.32	0.33	0.28	0.38	0.33	0.37	0.38

Content of p-MNT in EMSIT 1 stored in glass weighing bottles (%)

Temperature (°C)	20				40			
Time (months)	0	3	6	18	0	3	6	18
Open	0.45	0.38	0.29	trace	0.45	0.35	0.25	trace
Covered with PE foil	0.45	0.39	0.38	0.06	0.45	0.38	0.29	trace
Closed by ground lid	0.45	0.49	0.50	0.18	0.45	0.36	0.31	trace

indicating that the homogenous distribution of taggants in the explosives did not change during storage.

Results from the laboratory tests are shown in Table 5.1. The evaluation of these results has to take into account the fact that the release of taggant vapor from an open surface of an emulsion explosive, can be influenced by changes in emulsion structure in the surface layer. A glassy layer was observed on the surface of EMSIT 1 in the exposed weighing bottles. This layer was probably formed in the process of water evaporation, followed by changes in the physical structure of the emulsion. This phenomenon could be the reason for the relatively slow decrease of the taggant concentration with time [190].

In situations where post-blast examinations have to be carried out in a medium such as sea-water, it is of interest to determine the solubility of taggants in sea-water. Laboratory measurements produced the following results: 1.1×10^{-2}, 1.3×10^{-4}, 1.5×10^{-4}, and 2.5×10^{-3} g/100 g salt water for EGDN, *o*-MNT, *p*-MNT and DMNB, respectively [38]. For fresh water the solubilities for the same compounds are: 0.5, 0.05, 0.05, and 0.002 g/100 g fresh water, respectively. The thermal properties of the four ICAO taggants were studied, using differential scanning calorimetry (DSC) and heat flux calorimetry (HFC) [191–193]. The requirements of a taggant, from the point of view of thermal stability, are that it should be compatible with all potential plastic explosives, easily mixed during the manufacturing process, and not separate from, vaporize from, or react with the explosive during manufacture, storage or transportation, that is, neither thermal events nor 'aging problems'.

Some of the general properties of the taggants are summarized in Table 5.2. Since the taggant has to be added during the manufacture of the plastic

Table 5.2 General properties of taggants. From Jones, D. E. G. et al., *J. of Thermal Analysis*, 1995 **44**, 533. Copyright John Wiley & Sons Ltd. Reproduced with permission

Taggant	State (at 25°C)	Vapor pressure at 25°C (torr)	Impact sensitivity[a] H_{50} (cm)	Weight % in explosive[b]
EGDN	liquid	0.06	20	0.2
DMNB	solid	1.67×10^{-3}	>10	0.1
o-MNT	liquid	0.11	5	0.5
p-MNT	solid[c]	0.03	5	0.5

a: H_{50} (TNT) = 30 cm; b: as recommended by the ICAO; c: melts at 50°C.

explosives, DMNB has the advantage of being the only taggant that remains a solid at typical processing temperatures, i.e. >80°C. Compatibility of the taggant with the explosive in question is an important criterion for establishing the utility of a particular taggant. Thermal and mechanical properties of the taggant–explosive mixture have to be compared with the same properties for the individual taggant and explosive. The compatibility of the taggants with the explosive tetryl has been tested using DSC measurements. For all the cases, except that for the mixture containing p-MNT, the onset of the decomposition of tetryl was unaffected by the presence of the taggant. In the mixture containing p-MNT and tetryl, evidence for incompatibility was found because of a lower onset temperature than for pure tetryl and a failure of the mixture to adhere to an energy balance (in contrast to other mixtures of taggant + tetryl).

Tests made with all four taggants showed that only two taggants (DMNB and p-MNT) could be effectively used [194]. EGDN was too sensitive to various impulses, and could therefore hardly be added to plastic explosives, when taking into account the safety of production and use. o-MNT had a too high vapor pressure and its smell was unacceptable.

In order to improve the detection lifetime of sheet explosives, DMNB was microencapsulated with Parylene C at levels of 1–6% [195]. Thermogravimetric analysis (TGA) showed that the emission rate of the microencapsulated DMNB at any given temperature was less than that of untreated DMNB. It was estimated that the lifetime for detection would improve by at least a factor of two.

A feasibility study of incorporating DMNB into detonating cord has been carried out [196]. Thermal studies of mixtures of DMNB with PETN (the explosive ingredient of detonating cord), using DSC, TGA, and accelerating rate calorimetry (ARC), indicated that DMNB caused a small decrease in the thermal stability of PETN. Production trials with two types of detonating cord, Cordex-25 (polyethylene + textile/wax coating) and B-Line (PVC covering), containing about 5 g/m of PETN, with 0.05, 0.1, and 0.5% DMNB were carried out, showing no effect on the explosive performance of the cords. The type of cover on the detonating cord was found to have an effect on the emission rate of DMNB from the cord. The DMNB was depleted slower when polyethylene/fabric was the cover, as opposed to PVC, although the latter appeared to have absorbed some DMNB which became a source of vapor over time. Tests with marked detonating cord in standard ICAO suitcases and real suitcases, indicated that the detection characteristics were comparable to marked sheet and plastic explosives.

There is a great interest in marking emulsion explosives because they do not contain any volatile components which can be used for detection [194]. Initial studies used DMNB and p-MNT for marking emulsion explosives. Results showed that the emission rate of taggants from unpacked emulsion explosives was similar to that of taggants from plastic explosives, that is 10^{-5} to 10^{-6} g/cm^2·h. However, the multilayer packing film designed to stop the diffusion of water from cartridged emulsion explosives, can also stop the diffusion of taggants from these explosives.

Another idea which has come up is to limit the shelf-life of plastic explosives, in order to minimize their potential use for criminal purposes [194]. The principle could be based on certain rheological properties of plastic explosives which would cause a quick loss of plasticity after 1–1.5 years at room temperature. This loss of plasticity can be caused by cross-linking the polymer matrix (hardening of the plastic explosive) or by degradation of the binder. This method can be useful only for industrial plastic explosives where a shelf-life longer than one year is not necessary. Longer shelf-life could result in illegal use of such explosives.

Possible processes which could lead to the required changes in explosive plasticity have been suggested [194]:

- Creation of a suitable number of cross-links in the binder polymer.
- Degradation of the polymer by oxidation or photo-oxidation, which will lead to a substantial decrease in molecular weight to a value at which the polymer will lose its binding properties.
- Biodegradation of the binder, which requires the use of biodegradable polymers, such as polyhydroxypropionate, polyhydroxyvalerate, polyhydroxy-butyrate and copolymers of these compounds.

The method chosen for further experiments was controlled vulcanization of ethylenepropylene terpolymer in the binder. The velocity of vulcanization was checked in plastic explosives containing RDX and PETN. Various vulcanization additives were tested in order to find those leading to the required effect, that is quick vulcanization after 1–1.5 years.

Ammonium nitrate (AN) is produced in large quantities for use both as a fertilizer and as an ingredient in explosives. Considerable effort has been invested in reducing the effectiveness of fertilizer-grade AN as an explosive or to render it inert. In principle, explosive chemicals might be rendered inert by adding a chemical suppressant or diluent or by changing the physical form of the explosive [185]. Alternately, energetic materials might be desensitized to reduce their explosive power, or make them more difficult to detonate.

One such method claims to render fertilizer-grade AN resistant to flame and insensitive to detonation [197], by adding 5–10% of mono- and diammonium sulfate. Tests showed that AN containing the additives were detonable, when tested in sufficiently large charge diameters [185].

5.4 VAPOR DETECTION OF TAGGED EXPLOSIVES

Detection of tagged explosives is carried out by commercially available vapor detectors, using methods such as GC–ECD, GC–Chemiluminescence (GC–TEA), and IMS.

Detection experiments were made with tagged Seismoplast, a German plastic explosive [186]. The vapor detector used was a GC–TEA system. Figure 5.5 shows the calibration of the instrument with six explosives and three taggants. The explosive was put into a polythene bag, wrapped into cotton-wool and packed with newspapers in a paper box. Measurements were performed after several hours up to three days, measuring the intensity of the tag signal. The o-MNT taggant, at a concentration of 0.5% in the explosive, could already be

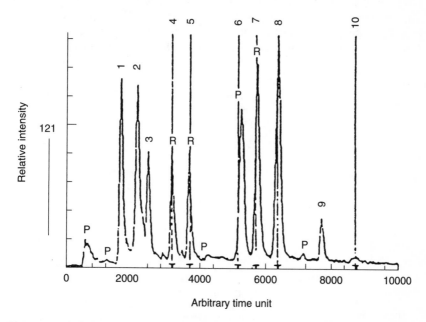

Figure 5.5 Calibration of GC–TEA system with explosives and taggants: 1, EGDN; 2, DMNB (A); 3, NG; 4, PETN; 5, RDX; 6, o-MNT; 7, p-MNT; 8, DMNB (B); 9, DNT; 10, TNT. Reproduced from Kolla, P., First Int. Symp. on Explosive Detection Technology, Atlantic City, NJ, 1991, p. 723, with permission.

detected after 1 h, while under the same conditions, *p*-MNT could be detected only after 3 h. The DMNB taggant, at a concentration of 0.5%, could be detected only after 24 h, and at a concentration of 0.1%, after 2 days.

Stability of the taggants in the plastic explosive, after one year of storage, was investigated. The explosives were shaped into small balls and put in a well ventilated place, without wrapping. After one year, samples were taken from the center of the balls and from the outer surface, and analyzed by HPLC and GC/MS in to measure the remaining amount of tagging compound. The DMNB-tagged explosive showed twice the concentration of DMNB at the center of the ball as at the outer surface. This was explained as a slow diffusion rate relative to the evaporation rate of the surface. Such an effect could largely reduce the detectability of that tagged explosive. GC/MS analysis of the explosives tagged with o- and *p*-MNT did not show any of the two taggants. It was concluded that the mononitrotoluene taggants are lost after one year of open storage.

A GC–ECD vapor detector was tested for detection of taggants [198]. The detector consisted of a two-component system: a battery-powered hand-held sampling unit and an analyzing unit (Figure 5.6). The analyzer unit consists of a desorber, dual chromatographic columns and an ECD detector. The sample collection tube (containing the adsorbed vapor or collected particulates) is placed in the continuously heated desorption unit of the analyzer. Desorption time is about 2–3 s, under a purge of pure nitrogen. The vapor sample then enters into the second adsorber, which serves to reduce sample contamination before introduction into the analytical columns. Separation of both volatile and

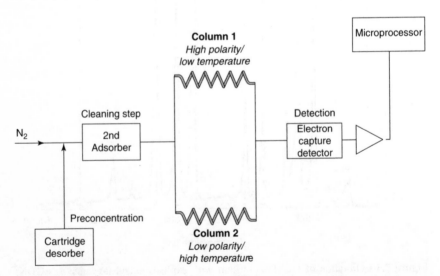

Figure 5.6 Block diagram of GC–ECD vapor detector for taggants. Reproduced from Nacson, S., et al., First Int. Symp. on Explosive Detection Technology, Atlantic City, NJ, 1991, p. 714, with permission.

relatively nonvolatile molecules takes place on the dual analytical columns and is detected by a single ECD detector.

Direct air sniffing of the head-space vapor of the taggants resulted in approximate detection limits of 5 p.p.t. for DMNB and 1 p.p.t. for EGDN and the mononitrotoluenes. Total analysis time per sample was 60 s, which included adsorption of the sample, thermal desorption, a precleaning step, chromatographic separation, detection, and data processing and reporting. A GC–ECD vapor detector was used to evaluate its sensitivity for the detection of DMNB in Composition C-4 [199]. C-4 containing 1.0 and 0.1% DMNB was prepared. The detector was capable of detecting about 5 pg of the taggant, with a detection time of 2 min. However, the detector was found to be nonlinear. Figure 5.7 shows the calibration curve for DMNB. The linear range extends only from about 100–400 pg (see the insert).

Detection of tagged explosives was done with a battery-operated Thermo-Redox handheld explosives vapor detector [38], which is schematically shown in Figure 5.8. The detector can be operated in either vapor or particulate mode. Sampling time is 5–30 s and analysis time 10 s.

The sensitivity of the detector for EGDN and DMNB at concentrations of 1–12 p.p.b. is shown in Figure 5.9. The lowest detection limit for these taggants, at room temperature, was 1 p.p.b. Assuming an emission rate of 483 p.p.b. of EGDN (0.2%) from tagged Semtex 1A [38, 189] and a factor of

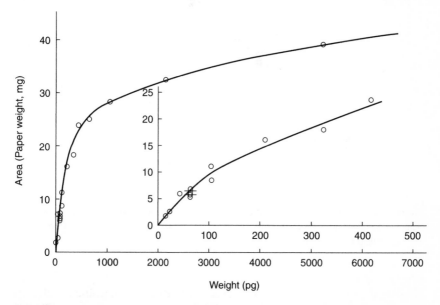

Figure 5.7 Calibration curve for DMNB. Reproduced from Chen, T. H., et al., First Int. Symp. on Explosive Detection Technology, Atlantic City, NJ, 1991, p. 741, with permission.

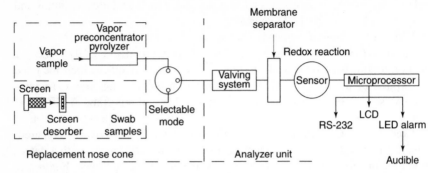

Figure 5.8 Schematic diagram of Thermo-Redox explosives vapor detector. Reproduced from Nacson, S., et al., Second Explosives Detection Technology Symp., Atlantic City, NJ, 1996, p. 38, with permission.

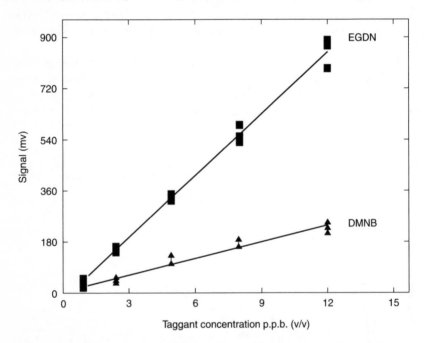

Figure 5.9 Detector sensitivity for EGDN and DMNB. Reproduced from Nacson, S., et al., Second Explosives Detection Technology Symp., Atlantic City, NJ, 1996, p. 38, with permission.

100 for vapor attenuation due to packaging and other barriers, EGDN can still be detected without difficulty. However, DMNB, at a concentration of 0.1% in Semtex A1, was below the 1 p.p.b. level. Higher concentrations of DMNB and/or improvements in the detector sensitivity would allow its detection with the Thermo-Redox detector.

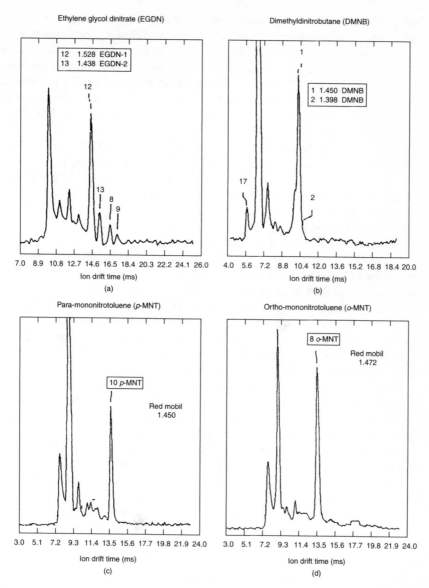

Figure 5.10 IMS spectra of taggants. Reproduced from Danylewych-May, L. L., et al., Advances in Analysis and Detection of Explosives, Dordrecht, 1993, p. 385, with kind permission from Kluwer Academic Publishers.

A study of taggants was performed with ion mobility spectrometry (IMS) [76]. The taggants studied were EGDN, *o*-MNT, *p*-MNT, and DMNB. IMS spectra of the taggants are shown in Figure 5.10. The inferred chemical identity of the ions formed was based on previously identified ions in the literature, using IMS/MS, and by enhancing peak intensity, by changing the reagent ion in the reaction chamber. The drift times at different temperatures were normalized to the drift time of $(TNT-H)^-$ at 100°C. Reduced mobilities were also calculated, based on measured drift times and assumed reduced mobility for $(TNT-H)^-$ to be $1.450\,cm^2/V\cdot s$. This value of reduced mobility was arrived at from calculations at high temperature (260°C) and assuming the values of 2.44 and $2.74\,cm^2/V\cdot s$ for the characteristic reduced mobilities of NO_3^- and Cl^-, respectively.

At an operating temperature of 100°C, the taggants EGDN and DMNB, could not be detected. By lowering the IMS operating temperature to 50°C, the two taggants were easily detected. The two peaks observed for EGDN (Figure 5.10a) are associated with chloride and nitrate ion attachment to the EGDN molecule. The intensity of the $EGDN\bullet NO_3^-$ ion is about one third of that of the $EGDN\bullet Cl^-$ ion, when pure EGDN is used. Traces of NG (peaks 8 and 9) are also observed in the IMS spectrum.

Increasing the concentration of the NO_3^- ion in the reaction chamber, enhances $EGDN\bullet NO_3^-$ formation. However, replacing the reagent ion Cl^- with NO_3^-, resulted in almost complete disappearance of both EGDN ions.

The ions observed in the IMS spectrum DMNB (Figure 5.10b) are formed by proton abstraction, and chloride and nitrate ion attachment. The effect of the NO_3^- ion concentration in the reaction chamber was similar to that of EGDN. The taggants *p*-MNT (Figure 5.10c) and *o*-MNT (Figure 5.10d) are detected readily by chloride ion attachment at 50°C.

Table 5.3 shows the reduced mobilities of the taggants. Table 5.4 shows the separate effects of reaction chamber length and drift tube temperature on the detection limits of the taggants. Significant sensitivity improvements for EGDN and DMNB are observed at lower temperatures and in the longer reaction chamber. The sensitivity of *o*-MNT and *p*-MNT did not improve in the longer reaction chamber at 100°C but improved at 50°C.

Gas chromatography with a surface acoustic wave detector (GC–SAW) was used for the detection of taggants [200]. The SAW detector (see paragraph 2.4.3) has a dynamic range of 100 000 at a constant temperature. In order to

Table 5.3 Reduced mobility data of taggants $(cm^2\,V^{-1}\,s^{-1})$ Reproduced from Danylewich-May, L. L., et al., *Advances in Analysis and Detection of Explosives*, Dordrecht, 1993, p. 385, with kind permission from Kluwer Academic Publishers

Taggant	MW	$(M-H)^-$	$M\bullet Cl^-$	$M\bullet NO_3^-$
EGDN	152	—	1.528 M	1.438 m
DMNB	176	1.556 m	1.450 M	1.388 m
o-MNT	137	—	1.472 M	—
p-MNT	137	—	1.450M	—

m: minor; M: major.

Table 5.4 Effect of ionization/reaction chamber length and drift tube temperature on taggant detection limit. Reproduced from Danylewich-May, L. L., et al., *Advances in Analysis and Detection of Explosives*, Dordrecht, 1993, p. 385, with kind permission from Kluwer Academic Publishers

Drift tube temperature (°C)	Chamber dimensions	Detection limits (ng)			
		o-MNT	*p-MNT*	*DMNB*	*EGDN*
100	Standard	25	25	n.d	n.d
100	Long	25	25	25	n.d
50	Standard	—	—	5	5
50	Long	1	1	0.5	0.1

n.d. = not detected

expand the range of the detector, two different SAW detector temperatures were used in the experiments. For large concentrations ($\mu g/cm^3$) of saturated vapor, a temperature of 60°C was used. For lower concentrations (pg/cm^3), a detector temperature of 15°C was used. The difference in sensitivity was sufficient to allow the SAW detector to be used over a large dynamic range of taggant vapor concentrations: $1 pg/cm^3$–$10 \mu g/cm^3$. Sample preconcentration (for 5 s) was used in all experiments.

Vapors for calibrating the GC–SAW were produced for liquid taggants (*o*-MNT and *m*-MNT) and compared with common solvents (benzene, toluene, and xylene), using the tedlar bag method illustrated in Figure 5.11 [201]. Tedlar bags of 10 and 1 L capacity were fitted with a nitrogen fill valve and a septum injection port. Using a syringe to inject a small amount of liquid into the bag, a validated vapor standard, typically 1 or 100 p.p.m., was produced and used to measure the sensitivity of the system. For taggants which are solids at room temperature, such as DMNB and *p*-MNT, as well as explosives containing taggants, a sealed vial with a septum lid was used to create saturated headspace vapors. For nonvolatile compounds, such as RDX, a small 1 mg sample was placed between glass wool packing in a glass tube as shown in Figure 5.12. The glass tube was then placed within a hollow cylindrical heater and a metered amount of air passed through the tube. This allowed the material to be raised to an elevated temperature and thus generate a measurable amount of vapor for testing.

The taggants DMNB and *p*-MNT are solids at room temperature and were tested by placing approximately $1 cm^3$ of the taggant into the sample vial, which allowed the collection of headspace vapors using the Luer inlet of the system and a syringe needle.

In these tests, saturated headspace vapor was the concentration level available for each test. DMNB and *p*-MNT co-elute, when using a DB-624 column. Sampling the saturated headspace vapor of DMNB produced a frequency shift of 3000 Hz. The vapor pressure at 25°C is 2.67 p.p.m., hence the scale factor

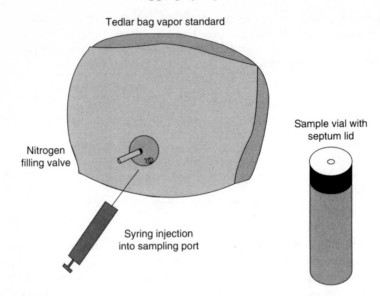

Figure 5.11 Tedlar bag method for production of vapors in GC–SAW. Reproduced from Staples, E. J., et al., First Annual Symp. on Enabling Technologies for Law Enforcement and Security, San Antonio, TX, 1996, with permission.

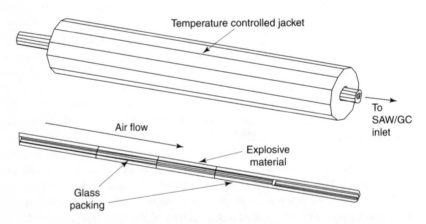

Figure 5.12 Production of vapors in GC–SAW from nonvolatile compounds. Reproduced from Staples, E. J., et al., First Annual Symp. on Enabling Technologies for Law Enforcement and Security, San Antonio, TX, 1996, with permission.

is 442 Hz/(p.p.m. cm^3). The response for *p*-MNT was between 3000–4000 Hz, with a 60°C detector temperature. The vapor pressure of *p*-MNT is 53 p.p.m. at 25°C, hence the scale factor is 120 Hz/(p.p.m. cm^3).

Minimal detection limits, using a 5 s preconcentration, for *o*-MNT, *m*-MNT, *p*-MNT, and DMNB were 11, 27, 6.7, and 18 p.p.b., respectively.

Saturated headspace vapors from Semtex, containing *p*-MNT, produced a response of 6000 Hz with a 60°C detector, with a very high signal-to-noise ratio. The response from Composition C-4 tagged with DMNB was lower, giving typically 3000–4000 Hz with a 15°C detector. Both tagged Semtex and C-4 could be detected in the low p.p.b. range, using a 5 s preconcentration (sample flow 45 cm^3/min). Retention times were slightly longer than 10 s. In order to achieve shorter retention times, a faster temperature ramp on the column would be needed.

6

Environmental detection
of explosives

6.1 INTRODUCTION

It has been known for many years that most explosives are toxic [11].
Therefore, they present a health hazard for ammunition workers and military
personnel who handle them and are exposed to them. During the first
$7\frac{1}{2}$ months of World War I, 17 000 TNT poisoning cases, including 475 deaths,
occurred in munition factories in the USA. However, in World War II,
during a period of 4 years, the total number of fatalities reported from all
government-owned explosive plants in the USA, amounted to 22. This lower
rate of explosive poisoning fatalities was believed to be the direct result of a
better understanding of the toxicity, diagnosis, and prevention of explosives
poisoning. During the 1960s, several cases of sudden death of workers in
dynamite production were reported in the USA and in Europe. Intoxication
with the RDX-based explosive composition C-4 occurred during and after the
Vietnam War.

The toxic effects of the most used explosives, TNT, RDX and NG, can be
summarized as follows [11]:

- TNT poisoning can lead to severe diseases such as aplastic anemia or toxic
  jaundice. In aplastic anemia, which is often fatal, the blood-forming organs
  fail to function, resulting in a progressive loss of the blood elements.
  Toxic jaundice indicates that the liver has been severely damaged. Less
  serious diseases resulting from TNT intoxication include cataract, dermatitis,
  gastritis, cyanosis, and symptoms of the nervous system. The EPA (USA
  Environmental Protection Agency) has determined that TNT is a possible
  human carcinogen. This assessment is based on a study in which rats that
  ate TNT for long periods developed tumors of the urinary bladder.
- Symptoms and clinical manifestations due to RDX poisoning include convulsions followed by loss of consciousness (which in some cases is followed

by death), muscular cramps, dizziness, headache, nausea, and vomiting. The EPA has determined that RDX is a possible human carcinogen.

- The most common symptoms of NG poisoning are mainly due to reduction of blood pressure. They include headache, throbbing in the head, palpitation of the heart, nausea, vomiting, and flushing. In severe cases, the heart muscle is affected directly and the heart beats are weakened. Death may occur from asphyxia due to paralysis of the respiratory center. Mental disturbances due to NG exposure include drowsiness, stupor, insomnia, mental confusion, dizziness, hallucinations, and maniacal manifestations.

Because of their toxicity, explosives not only present a health hazard to munition workers and military personnel who handle them, but they also constitute a general environmental problem. The disposal of large quantities of explosives in an environmentally acceptable manner poses serious difficulties. For many years, obsolete explosives and munitions were buried in the ground or dumped in the sea. Explosives-containing wastewater from explosives and munition-manufacturing plants and their degradation products used to be discharged into rivers and streams. A single TNT-manufacturing plant can generate as much as 500 000 gallons of wastewater per day, which contains TNT and other nitro compounds. Other methods used for disposal of explosives, such as open burning or detonation, are not acceptable today, because of air pollution by particulates, nitrogen oxides, and other toxic products. Methods of incineration, which overcome some of these difficulties, are available, but are very costly. Biological disposal of explosives is being investigated, but so far no practical method has been developed.

In order to assess the extent of explosives pollution of a contaminated area, it is necessary to detect and identify the explosives and their degradation products in groundwater and soil. This can be done by collecting soil and water samples and bring them to the laboratory for analysis by a variety of analytical methods [1].

However, there is much interest in detecting the explosives on-site in the contaminated area, using field tests or mobile detectors. Fast screening is especially important for the determination of explosives in soil before they decompose and leach through the water-unsaturated zone into groundwater.

For example, the Lifetime Health Advisory values for TNT and RDX in drinking water have been set at $2 \mu g/L$ (2 p.p.b.) [2, 12].

Some of the analytical methods, such as GC, HPLC, GC/MS, IMS, etc., have been adapted for use in field tests in the contaminated sites. However, this chapter will deal only with field tests and explosive detectors specially developed for environmental detection and/or screening of explosives on-site.

6.2 DEGRADATION OF EXPLOSIVES

Before describing the various methods available for environmental detection of explosives, it important to understand the main degradation processes occurring in explosives, and to recognize the presently known degradation products. The following is a short outline of the degradation processes and products of the main explosives.

6.2.1 Degradation products of TNT

Exposure of TNT, both as a solid and in solution, to strong sunlight or ultraviolet radiation, results in the formation of decomposition products. As a result of photodecomposition, aqueous solutions of TNT first turn pink, then gradually after a period of 4–6 h, change into a rusty-orange colored solution, known as 'pink water'.

Irradiation of an aqueous solution of TNT with a mercury arc lamp, fitted with a Pyrex filter so that only light above 280 nm passed, resulted into the following photodecomposition products:

1,3,5-trinitrobenzene (TNB), 4,6-dinitroanthranil, 2,4,6-trinitrobenzaldehyde, and 2,4,6-trinitrobenzonitrile [202].

This radiation is similar to solar radiation, because the earth's atmosphere absorbs 99% of the radiation below 280 nm emitted by the sun.

In a second experiment, an aqueous solution of TNT was allowed to stand in sunlight for 4 days. After this time, four new compounds, in addition to those found with the UV lamp, were identified:

2,2′,6,6′-tetranitro-4,4′-azoxytoluene (4,4′-Az),
4,4′,6,6′-tetranitro-2,2′-azoxytoluene (2,2′-Az),
2′,4-dimethyl-3,3′,5,5′-tetranitro-ONN-azoxybenzene, and
2,4′-dimethyl-3,3′,5,5′-tetranitro-ONN-azoxybenzene.

Components found in 'pink water', formed by irradiation of an aqueous solution of TNT with UV light, included:

1,3,5-trinitrobenzene (TNB),
4,6-dinitroisoanthranil,
4,6-dinitroanthranil,
2,4,6-trinitrobezaldehyde,
2,4,6-trinitrobenzonitrile,
2,4,6-trinitrobenzaldoxime,
2,4,6-trinitrobenzyl alcohol,
3,5-dinitrophenol,
2-amino-4,6-dinitrobenzoic acid,
2,2′-dicarboxy-3,3′,5,5′-tetranitroazoxybenzene,
2,2′-dicarboxy-3,3′,5,5′- tetranitroazobenzene,
2-carboxy-3,3′,5,5′-tetranitroazoxybenzene,
2,4,6-trinitrobenzoic acid, and
N-(2-carboxy-3,5-dinitrophenyl)-2,4,6-trinitrobenzamide.

Microbial degradation has been found to be a major source of formation of degradation products in TNT. It has been suggested that the reduction of nitro groups to amino groups proceeds through the nitroso and hydroxylamino compounds according to the following equations [203]:

$$R–NO_2 \underline{\quad} H_2 \longrightarrow R–NO + H_2O \qquad (6.1)$$

$$R–NO \underline{\quad} H_2 \longrightarrow R–NHOH \qquad (6.2)$$

$$R–NHOH \underline{\quad} H_2 \longrightarrow R–NH_2 + H_2O \qquad (6.3)$$

A suggested biotransformation scheme is shown in Figure 6.1 [204]. The main degradation products of TNT found in contaminated soils are:

2-amino-4,6-dinitrotoluene (2-A),
4-amino-2,6-dinitrotoluene (4-A),
2,4-diamino-6-nitrotoluene (2,4-DA), and
2,6-diamino-4-nitotoluene (2,6-DA).

The manufacture of TNT includes treatment of liquid toluene with mixed nitric and sulfuric acids. Following the nitration, the undesired TNT isomers and the residual DNT isomers are removed by conversion into soluble species and by extraction [13]. It is therefore expected that in areas contaminated by wastewater from TNT manufacture, isomers of DNT and TNT will be found in addition to amino compounds.

6.2.2 Degradation products of tetryl

Tetryl is often found together with its hydrolysis product, *N*-methyl picramide [205–207]. This thermal degradation product is formed in the presence of traces of water [208].

Photolytic products of tetryl in aqueous solution were found to include picrate ion, *N*-methylpicramide,methyl nitramine, nitrate and nitrite [209].

By-product impurities from production of tetryl were reported to include [210]:

N-methyl-*N*,2,4-trinitroaniline,
picryl chloride,
picric acid,
N,2,4,6-tetranitroaniline, and
N-methyl-2,4,6-trinitroaniline.

6.2.3 Degradation products of RDX

Biodegradation of RDX was found to occur under anaerobic conditions, yielding a number of products, including [211]:

hexahydro-1-nitroso-3,5-dinitro-1,3,5-triazine (MNX),
hexahydro-1,3-dinitroso-5-nitro-1,3,5-triazine (DNX), and
hexahydro-1,3,5-trinitroso-1,3,5-triazine (TNX).

A scheme for the biodegradation of RDX (Figure 6.2) was proposed which proceeded via successive reduction of the nitro groups to a point where destabilization and fragmentation of the ring occurred. The noncyclic degradation

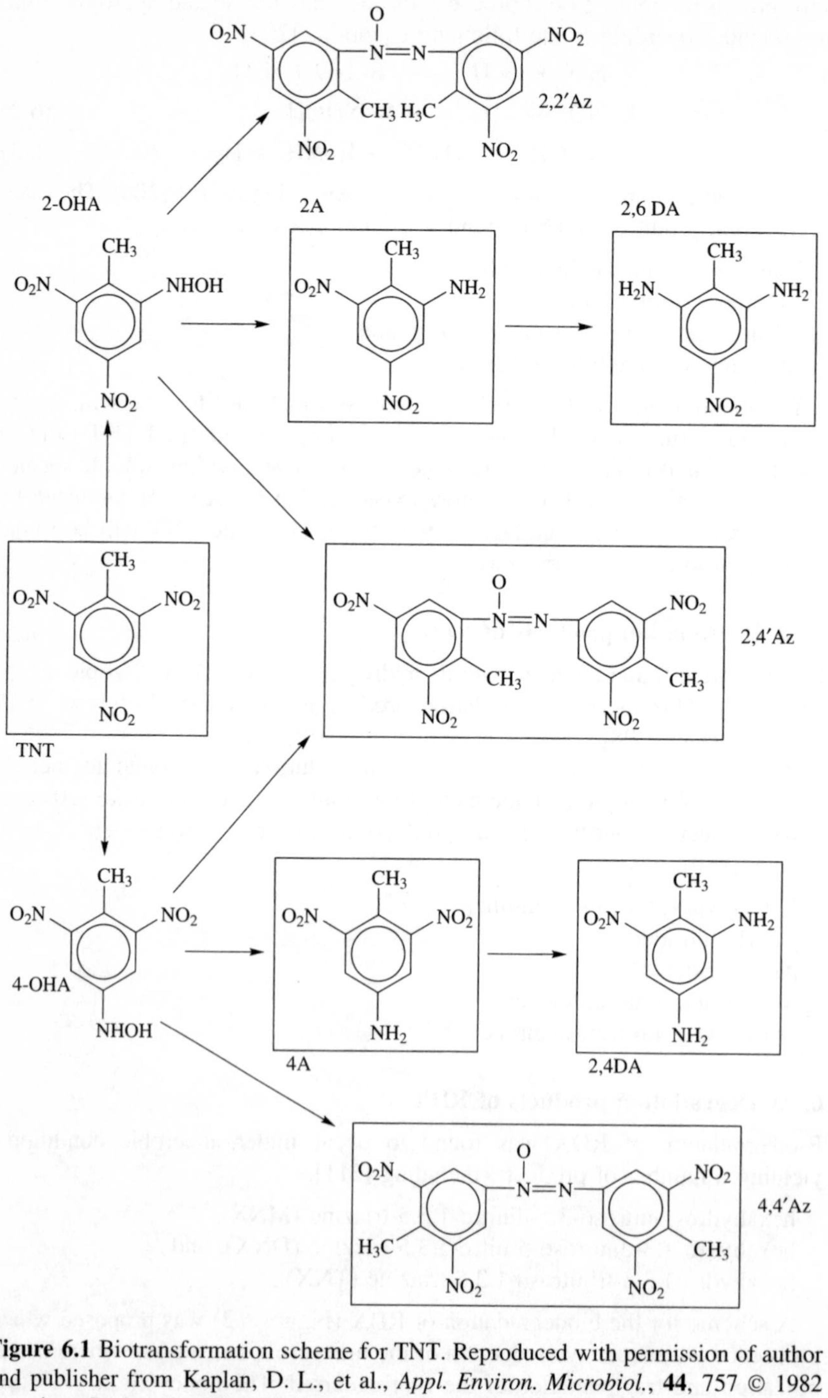

Figure 6.1 Biotransformation scheme for TNT. Reproduced with permission of author and publisher from Kaplan, D. L., et al., *Appl. Environ. Microbiol.*, **44**, 757 © 1982 American Society for Microbiology.

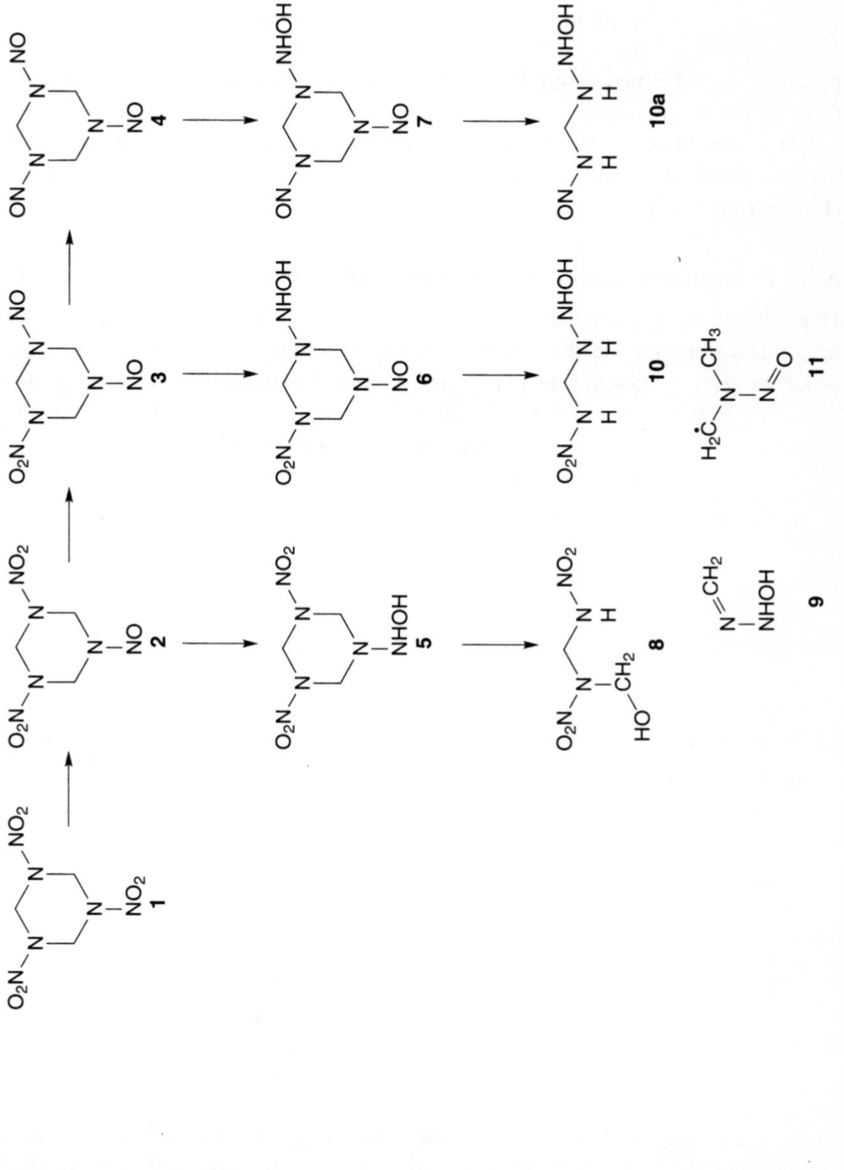

Figure 6.2 Biodegradation scheme of RDX. Reproduced with permission of author and publisher from McCormick, N. G., et al., *Appl. Environ. Microbiol.*, **42**, 817 © 1981 American Society for Microbiology.

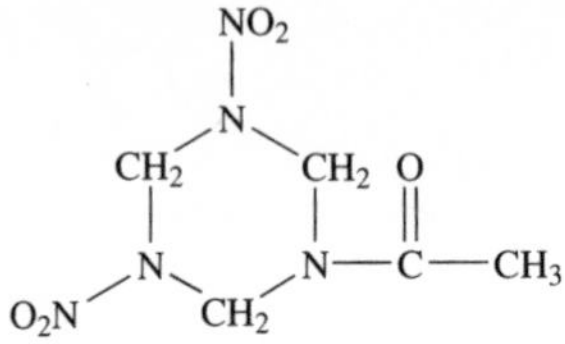

NO$_2$
|
N
CH$_2$ CH$_2$ O
| | ||
N   N—C—CH$_3$
O$_2$N   CH$_2$

Figure 6.3 Chemical structure of TAX.

products were formed through subsequent reduction and rearrangement reactions of the fragments.

RDX wastewater from manufacturing plants contains also by-products of the manufacturing process. The most abundant by-product of RDX is TAX (Figure 6.3) [212].

6.2.4 Degradation products of nitrate ester explosives

Degradation of glycerol trinitrate (nitroglycerin, NG) has been reported to take place stepwise via the dinitrate and mononitrate isomers, with each step proceeding at a slower rate [213]. These decomposition products were also found in post-explosion extracts when NG was involved in the explosion. Possible biodegradation pathways of NG are shown in Figure 6.4.

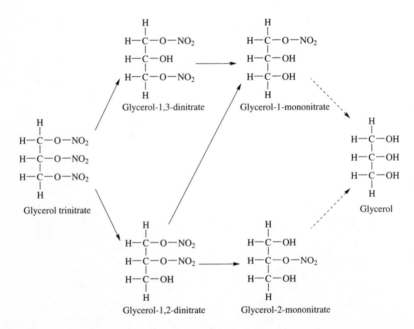

Figure 6.4 Biodegradation pathways of NG. Reproduced with permission of author and publisher from Wendt, T. M., et al., *Appl. Environ. Microbiol.*, **36**, 693, © 1978 American Society for Microbiology.

Lower nitrate esters of pentaerythritol, probably due to hydrolytic decomposition processes, were found in post-explosion debris, when PETN was involved in the explosion [214]. They included pentaerythritol trinitrate, pentaerythritol dinitrate and pentaerythritol mononitrate.

Similar degradation products were observed in the metabolic pathways of PETN [11].

6.3 CHEMICAL COLOR SCREENING TESTS

A rapid on-site device for the detection of TNT in effluent water from the purification apparatus of ammunition plants, based on ion-exchange resins, was developed [215]. The apparatus consists of an indicator tube, a fitting, and a syringe (or syringe pump). The indicator tube consists of a glass tube (4 mm inner diameter, 6 mm outer diameter, 10 cm long), containing a basic oxide section and an anion exchange resin indicator section, separated and held in place by glass wool plugs. Best results were obtained with an alkyl quaternary amine chloride exchange resin that was lightly colored (Dowex 1 × 10). Basic oxides, strong enough to convert TNT into its Meisenheimer colored anions [5], included NaOH, CaO, and MgO. CaO-coated glass beads were used. The Meisenheimer anions displaced the Cl^- anions of the resin to become strongly attached to the cations of the anion-exchange resin, forming a reddish discoloration of the resin. The length of the discoloration on the resin was proportional to the concentration of TNT in the wastewater tested. By comparison of the stain length from an analyzed sample with that obtained from standards, the concentrations of the unknown sample could be estimated. The detection limit of this method for TNT in water was 100 p.p.b.

A membrane was developed for the direct determination of TNT in groundwater [216]. A typical membrane is prepared by dissolving the following compounds in tetrahydrofuran: 0.5 g of poly(vinyl chloride) (PVC), 0.2 mL of dioctyl phthalate (DOP) to serve as plasticizer and 0.12 mL of Jeffamine T403, a polyoxyethyleneamine which reacts with TNT to produce a colored product. The membrane is formed by casting the solution into a 8 cm diameter glass Petri dish, and allowing the solvent to evaporate slowly. Detection limits of 10 ng/mL of TNT in water were obtained, by allowing the membrane to react with the sample for enough time for colored product to accumulate. The reaction was found to be irreversible, therefore, the membrane could be used only for sensing on an integrating basis.

Because the response to TNT was found to be independent of membrane thickness, it was believed that the initial color-forming reaction occurred at the membrane surface, rather than being uniformly distributed through the membrane [217].

In addition to 2,4,6-TNT, the membrane was also found to react with 2,4,5-TNT, tetryl, and 1,3,5-trinitrobenzene, to form colored products. However, the absorption spectra of the products differ for different explosives. The membrane does not react with RDX and HMX to form a colored product. With 2-amino-4,6-dinitrotoluene, a light yellow color was observed.

Dioctyl phthalate (DOP), which serves as a plasticizer, keeps the membrane soft and elastic, which facilitates its handling. The presence of DOP in the membrane increases the rate of diffusion of both TNT and amine, leading to a faster response, and may also help keeping the amine in the membrane while excluding water.

The membranes remain clear indefinitely in water. After storage for 80 days in air in a sealed container, over 80% of the amine remains in the membranes. After exposure to water for 10 days, over 40% of the amine remains in the membranes, which continue to respond to TNT. The membrane can be used to determine TNT in untreated water samples of pH 6–9. Recoveries of 0.1–4.0 mg/L from spiked groundwater ranged from 95 to 105%.

The membrane described above was adapted for remote measurements through a single optical fiber [218]. Refractive index matching to reduce stray light and a reflector behind the membrane to increase the reflected intensity kept the stray levels small relative to the signal of interest. It was estimated that TNT levels as low as 0.10 mg/L could be determined by this technique.

PVC membranes and thin films containing Jeffamine T-403 have been used to detect 2,4-DNT and TNT vapors [219]. Polymer membranes 100–150 µm thick and 10–20 µm thin films were used. Polymer membranes were exposed to either 2,4-DNT or TNT vapors for several days, at a temperature of 50°C. The membranes had a peak absorbance of approximately 500 nm after exposure to TNT vapors and a peak absorbance of approximately 430 nm after exposure to DNT vapors. These were found to be spectrally similar to those of the membranes used for detection of explosives in groundwater, suggesting that the reaction products are also chemically similar. Membrane response to DNT vapor was found to be much faster than TNT as could be seen by a faster color formation rate. Also, the lower vapor pressure of TNT and its inherent nature to attach to surfaces including soil particles, might limit the practical use of this sensor to detect TNT vapor directly. The thin films were studied for their response to DNT vapors and showed faster response at ambient temperatures.

In situ optical detection of explosives based on quenching of fluorescence by nitro compounds has been reported [220]. A fluorophore was incorporated into a plasticized polymeric membrane, which extracted organic nitro compounds from water. The membrane responded reversibly to all organic nitro compounds, including nitramines. Membranes were prepared by solvent casting. 0.2–0.3 g polymer, 0.1–0.2 mL plasticizer and fluorophore were dissolved in 6–8 mL of cyclohexanone. After mixing, the solution was poured into a 8 cm diameter glass Petri dish. The solvent was allowed to evaporate, which took 24–72 h. The resulting membrane was cut into small sections, each weighing 2–5 mg and having an area of about 0.5–0.6 cm^2. The membrane thickness was about 0.08 mm. The most sensitive membrane was prepared by using pyrenebutyric acid (PBA) as fluorophore, cellulose triacetate as polymer, and isodecyl diphenyl-phosphate (IDP) as plasticizer. A plasticizer must be added in order to obtain membranes that are clear and flexible. The plasticizer was also found to have a strong influence on the ability of the membrane to extract RDX from aqueous solutions. For a given set of conditions, the primary factor determining the sensitivity is the extent to which each nitro compound partitions into the membrane. Detection limits were found to be 2 mg/L for DNT and TNT, and 10 mg/L for RDX. Nitrogen purging before the measurement enhances the sensitivity and eliminates interference from oxygen.

A color test was developed for semiquantitative determination of explosives in water [221]. Explosives measured included TNT, DNT, RDX, and HMX. The method consists of concentrating aqueous solutions of explosives on a porous film adsorbent, followed by chemical color reactions using *o*-tolidine and Griess reagent [5]. Silica gel, HF254, was used as an adsorbent. The porous discs of silica gel were prepared by the following procedure: 15 mL of acetone and 1 mL of 10% cellulose acetate solution were added to 0.5 g of dried powdered silica. Filter paper was inserted into this suspension for 20 s. After drying, the complex obtained (adsorbent film + filter paper) was cut into discs 12 mm in diameter. By photometric measurement of the colored surfaces of the adsorbent, and using reference curves obtained by the same procedure, concentrations were determined.

Detection limits were estimated to be 0.2 mg/L for solutions of 50 mL. The method was unable to distinguish between RDX and HMX.

A color field test for the screening of TNT in soil was developed [222]. The test is based on the well-documented [5] reaction between polynitroaromatic compounds and alkalis, producing colored complexes known as Meisenheimer complexes. An amount of 6 g of soil is placed in a test tube 200 × 25 mm or larger, followed by the addition of 35 mL of methanol and vigorously shaken for about 1 min. Addition of 2 drops of 10% aqueous sodium hydroxide to 3 mL of the decantate will result in a pink to purple color, if TNT is present. The concentration of TNT can be determined by the absorbance at 516 nm, using a battery-operated spectrophotometer, or by comparison of the color with a set of standards prepared from site background soils and prepared as described above.

The detection limit was estimated to be in the range 4–8 p.p.m. The results of the screening tests were found to correlate well with the HPLC results. In another color field test for TNT, 2,4-DNT, and RDX, a 20 g portion of soil is extracted by manually shaking it with 100 mL of acetone for three minutes, and filtering the extract with disposable syringe filters [223]. After the soil settles, the decantate is divided into three aliquots. For the TNT test, a pellet of potassium hydroxide and about 0.2 g of sodium sulfite were added to 25 mL of acetone soil extract. The mixture was manually shaken for 3 min, then filtered. The presence of TNT will be indicated by the appearance of a red color. Absorbance was measured at 540 nm. A similar procedure is used for the 2,4-DNT test, except that two pellets of pottassium hydroxide and about 0.75 g of sodium sulfite were added, the samples were shaken for 1 min, allowed to stand for 28 min, and then shaken again for 1 min, before filtration. The presence of 2,4-DNT will be indicated by the appearance of a blue–purple color. Absorbance was measured at 570 nm. For the RDX test, the third aliquot of the extract is passed through a strong anion exchange cartridge at 5 mL/min to remove any nitrate and nitrite which could be present. A 5 mL aliquot was acidified with 0.5 mL glacial acetic acid and reacted with 0.3 g zinc dust in the barrel of a syringe fitted with a disposable filter unit. The solution was filtered into a vial containing 20 mL distilled water, to which a Griess reagent was added. The sample was shaken briefly and allowed to stand for 10—15 min. The presence of RDX will be indicated by the appearance of a pink color. Absorbance was measured at 507 nm. Figure 6.5 shows a flow diagram of the field method.

Concentrations of TNT, 2,4-DNT, and RDX were estimated from the absorbances of their colored products at 540, 570, and 507 nm, respectively. Detection

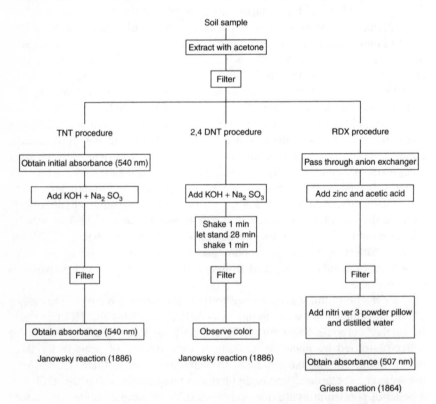

Figure 6.5 Flow diagram for colorimetric field methods for TNT, DNT and RDX. Reprinted from Jenkins, T. F., et al., *Talanta*, **39**, 419, © 1992, with permission from Elsevier Science.

Table 6.1 Colors and λ_{max} for acetone solutions of explosives and related compounds with: KOH and sodium sulfite; zinc and acetic acid followed by Griess reagent. Reprinted from Jenkins, T. F. et al., *Talanta* **39**, 419, © 1992, with permission from Elsevier Science

| Compound | KOH and Na$_2$SO$_3$ | | Zinc and acetic acid, Griess reagent | |
	Color observed	λ_{max} (400–600 nm)	Color observed	λ_{max} (400–600 nm)
1,3-Dinitrobenzene	Purple	570	None	—
2,4-Dinitrotoluene	Blue	570	None	—
2,6-Dinitrotoluene	Pinkish-purple	550	None	—
1,3,5-Trinitrobenzene	Red	460, 560	None	—
Tetryl	Orange	460, 550	Pink	507
2-Amino-DNT	Pale yellow	400	None	—
4-Amino-DNT	None	—	None	—
Nitroglycerin	None	—	Pink	507
PETN	None	—	Pink	507
RDX	None	—	Pink	507
HMX	None	—	Pink	507
Picric Acid	Reddish-orange	420	None	—
2,4-Dinitrophenol	Yellowish-orange	430	None	—
TNT	Red	462, 540	None	—

limits were about 1 µg/g for TNT and RDX, and 2 µg/g for 2,4-DNT. Heavy metal cations, such as copper, were found to interfere with 2,4-DNT determinations. The color-forming reactions used for these field screening tests were found not to be specific for TNT, 2,4-DNT, and RDX. Other polynitroaromatic compounds, such as 1,3-dinitrobenzene and polynitrophenols, such as picric acid, also produce colored anions when reacted with a strong base. Similarly, the same azo dye produced from the RDX test, is also produced when other nitramines, such as HMX and tetryl, or nitrate esters, such as NG, PETN and nitrocellulose (NC), are treated under similar conditions.

Table 6.1 shows a list of explosives and related nitro-compounds detected by these screening methods.

If 2,4-DNT is present as a minor component with TNT as a major one, 2,4-DNT will not be detected by this method. The absorbance for 2,4-DNT was found to be dependent on the water content, therefore, the screening test for 2,4-DNT is only semiquantitative.

6.4 IMMUNOCHEMICAL DETECTION METHODS

6.4.1 Introduction

Immunoassays are immunochemical detection methods based on a reaction between a target analyte and a specific antibody [224]. Quantitation is performed by monitoring a color change or by measuring radioactivity or fluorescence. Immunochemical methods rely on antibodies typically produced in rabbits, sheep, or goats for polyclonal preparations, or in mice or rats for monoclonal preparations. Antibody molecules are composed of equal numbers of heavy and light polypeptide chains, held together by disulfide bonds. Analytes of low molecular mass must be coupled to carriers such as proteins, to stimulate antibody production. Frequently, a chemical functionality, such as $-OH$, $-COOH$, $-NH_2$, or $-SH$, must be introduced onto the target analyte to form a hapten. The hapten is then covalently conjugated to a carrier molecule to form an antigen for the production of antibodies. The design and synthesis of the appropriate hapten and subsequent hapten–protein conjugate are major steps in the development of immunochemical methods for explosives. Structurally, the hapten must be closely related to the analyte, so that that the analyte will bind to the resulting antibodies.

One common immunoassay is the enzyme-linked immunosorbent assay (ELISA). The specificity of the antibody for the analyte and the resultant immune complex is the basis for the specificity of immunoassays. In a typical field-portable ELISA, a solid support (microtiter plate) is coated with a hapten–protein conjugate. An analyte-specific antibody is immobilized onto the microtiter support. A wash solution containing buffer removes any remaining unbound antibody. To perform an analysis, standards and samples are diluted with buffer and placed onto the microtiter plate. A constant, known amount of enzyme-labeled analyte is then added (alkaline phosphatase and horseradish are commonly used as labels). Competition between the analyte in the sample and the enzyme-labeled analyte occurs for the immobilized antibody. After a short incubation period, all unbound materials are removed by washing the plate with a phosphate buffer. An enzyme substrate and a chromogen are added, causing the development of a color. Quantitation of the target analyte is based on competition between the target analyte in the sample and enzyme-labeled analyte that was added. The greater the amount of analyte in the sample, the lighter the color produced at the end of the assay. Conversely, if samples contain only a small amount of the analyte, a dark color is produced because more enzyme-labeled analyte can bind to the immobilized antibody. Absorbance can be measured with a portable spectrophotometer or estimated visually.

6.4.2 Immunoassay methods for the detection of explosives

6.4.2.1 Enzyme-linked immunosorbent assays (ELISAs)

Several enzyme-linked immunosorbent assays (ELISAs) have been developed for TNT, 1,3,5-trinitrobenzene, 2,4-dinitrotoluene, and 2,6-dinitrotoluene, using polyclonal antibodies raised in New Zealand white rabbits [225].

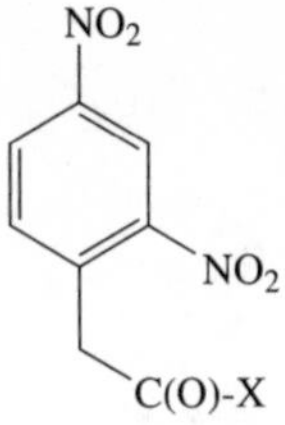

8a X = NHS
8b X = BSA
8c X = OVA

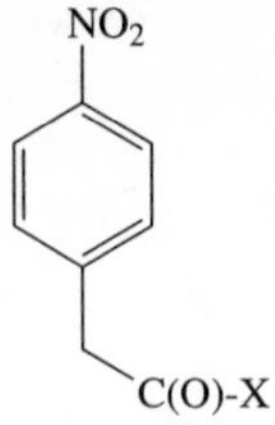

9a X = NHS
9b X = BSA
9c X = OVA

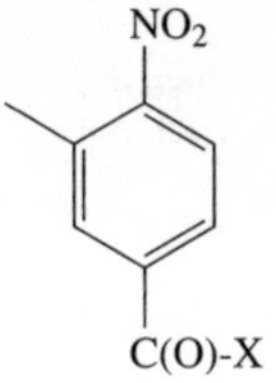

10a X = NHS
10b X = BSA
10c X = OVA

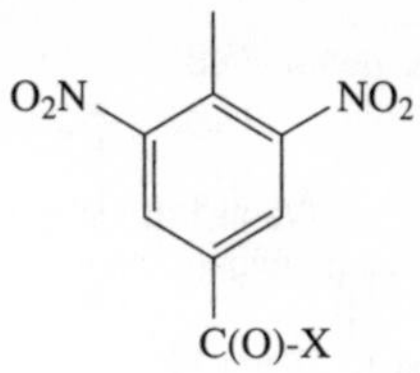

11a X = NHS
11b X = BSA
11c X = OVA

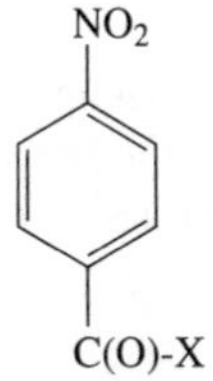

12a X = NHS
12b X = BSA
12c X = OVA

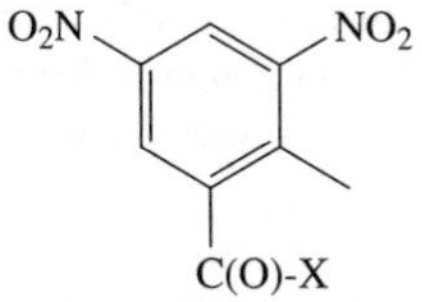

13a X = NHS
13b X = BSA
13c X = OVA

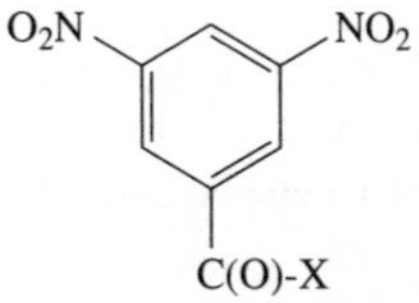

14a X = NHS
14b X = BSA
14c X = OVA

Figure 6.6 Structures of the NHS esters of selected nitroaromatic haptens, the hapten-BSA conjugates and the hapten-OVA conjugates. Reprinted with permission from *Immunochemical Methods for Environmental Analysis*, Vol. 42, American Chemical Society, Washington, DC, Chapter 9, © 1990 American Chemical Society.

Nitro substituted benzoic and phenyl acetic acids were used as haptens by conversion to the corresponding *N*-hydroxysuccinimide (NHS) esters, followed by coupling to protein carriers. In this study, haptens conjugated to bovine serum albumin (BSA) were used as immunizing antigens and hapten conjugated to chicken egg albumin (OVA) were used as coating antigens. Figure 6.6 shows the structures of the NHS esters of selected nitroaromatic haptens, the hapten–BSA conjugates and the hapten–OVA conjugates. Four compounds were selected to be injected into rabbits as immunizing antigens. For example, the BSA conjugate of 2,4-dinitrophenylacetic acid seemed a likely immunogen for 2,4-dinitrotoluene or 2,4,6-trinitrotoluene (TNT), while the conjugate of 3,5-dinitro-4-methylbenzoic acid was selected as a target for 2,6-dinitrotoluene. The antibodies, which were developed to 1,3-dinitroaromatic haptens, had the greatest specificity and sensitivity when the nitroaromatic analytes contained a 1,3-dinitro functionality. In one ELISA system, a lower detection limit for various 1,3-dinitroaromatic analytes of 1 ng/mL with an I_{50} (50% of the control) of 5 ng/mL was observed. Figure 6.7 shows the sensitivity of TNT to four different combinations of coating antigen

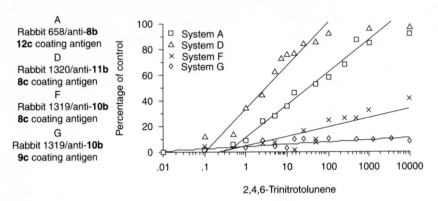

Figure 6.7 Sensitivity of TNT to four different combinations of coating antigen and antisera. Reproduced with permission from *Immunochemical Methods for Environmental Analysis*, Vol. 42, American Chemical Society, Washington, DC, Chapter 9, © 1990 American Chemical Society.

and antisera. Antisera from rabbit 1320/anti-**11b** with coating antigen **8c** (system D) shows excellent sensitivity for TNT.

No cross reactivity with mononitroaromatic compounds was observed. Antibodies, developed to mononitroaromatic haptens, showed high affinity for a variety of coating antigens but would not compete with nitroaromatic analytes in a normal ELISA.

An enzyme-linked immunosorbent assay (ELISA) for the determination of TNT and other nitroaromatic compounds was developed [226, 227]. As tracer, a trinitrophenyl-derivative conjugated to horseradishperoxidase, via 3-aminopropionic acid, was used. All reactions were performed on transparent polystyrene microtitre plates. The antibody solution was diluted $1:25\,000$ in 0.05 M sodium carbonate buffer pH 9.6 containing 3 mM sodium azide; 200 µL was employed in each well and incubated for at least 4 h at room temperature. The plates were washed four times with phosphate buffered saline (PBS) pH 7.6, containing 0.5% (v/v) polyoxyethylene sorbitan (Tween 20). Standard solutions or samples (200 µL/well) were pipetted into the wells and preincubated for 15 min. The peroxidase tracer was diluted $1:50\,000$ in PBS, pH 7.6, and 100 µL was added. After a further incubation of 15 min, the plate was washed as described above; 200 µL of a freshly prepared substrate solution (hydrogen peroxide and tetramethylbenzidine) was added. The reaction was stopped after another 15 min by addition of 100 µL of 5% sulfuric acid. The absorbance was then measured at a wavelength of 450 nm. A scheme of the ELISA procedure is shown in Figure 6.8.

TNT could be detected within the range 0.02–20 µg/L, with a detection limit of about 20 ng/L. Cross-reactivities with several nitroaromatic compounds were examined at different concentrations. Among the nitroaromatic tested compounds, 2,4-DNT had the highest cross-reactivity.

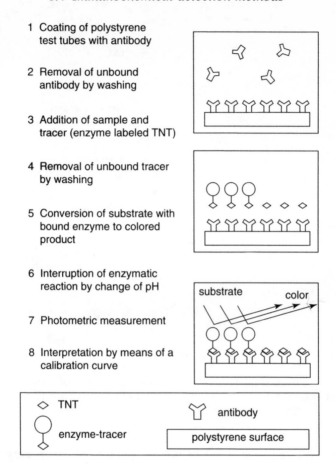

1 Coating of polystyrene
  test tubes with antibody

2 Removal of unbound
  antibody by washing

3 Addition of sample and
  tracer (enzyme labeled TNT)

4 Removal of unbound tracer
  by washing

5 Conversion of substrate with
  bound enzyme to colored
  product

6 Interruption of enzymatic
  reaction by change of pH

7 Photometric measurement

8 Interpretation by means of a
  calibration curve

Figure 6.8 Scheme of the ELISA procedure. Reproduced from Keuchel, C., et al., Anal. Sci., 8, 9 (1992). By permission of the Japan Society for Analytical Chemistry.

This method was the basis for a rapid field screening test for TNT in water and soil [228]. The immunofiltration assay is based on a simplified enzyme-linked immunosorbent assay (ELISA) performed in a pre-packed portable device. The tracer was prepared by conjugating the 6-aminohexanoic acid derivative of 2,4,6-trinitrophenyl to horseradish peroxidase by a *N*-hydroxysuccinimide/carbodiimide method [226]. The immunofiltration assay was performed in a white plastic device equipped with three different layers (Figure 6.9). A cotton pad, 10 mm thick, was placed in the lower part of the device. The second layer was a filter paper (20×20 mm) to achieve regular flow rates, on top of which was a porous membrane (20×20 mm), capable of adsorbing the antibodies. The layers were fixed with a plastic lid which had a central circular hole of 10 mm diameter.

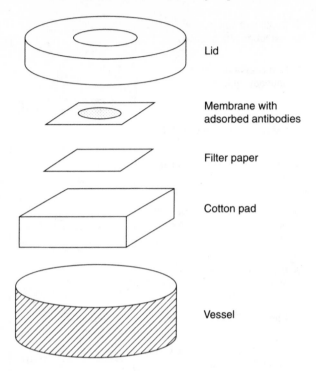

Figure 6.9 Construction of immunofiltration device. Reproduced with permission of author and publisher from Keuchel, C., et al., *Fresenius J. Anal Chem.*, **350**, 538 (1994). © 1994 Springer-Verlag Berlin.

All solutions were prepared just before use and diluted if necessary. Mouse antiserum and the TNT antibody solution were diluted 1 : 100 in phosphate buffered saline (0.08 mol/L phosphate, 0.145 mol/L NaCl, pH 7.6). The trinitrophenyl–peroxidase conjugate (TNP–POD) tracer was diluted 1 : 5000 in the same buffer. The assay of the peroxidase was performed with 3,3′,5,5′-tetramethylbenzidine/H_2O_2 in citrate buffer (0.2 mol/L citrate, 0.01% sorbic acid potassium salt, pH 3.8). Since a dye is needed, which is insoluble in aqueous buffers and precipitates on the membrane, an enhancer solution was added at a ratio of 1 : 10 (v/v). The following solutions were added to the upper membrane and allowed to drain through the membrane: 100 μL mouse antibody, 200 μL TNT antibody, 100 μL PBS buffer, 200 μL sample or standard, 100 μL tracer, 100 μL PBS solution, 200 μL substrate solution. After the color development, the reaction was stopped by the addition of 100 μL citrate buffer. After draining, the colored membrane was removed from the device and placed in a portable reflectometer, where the reflectance was measured at 573 nm. The reaction of peroxidase with the substrate leads to a blue colored charge transfer complex between reduced and oxidized tetramethylbenzidine. In the presence of the enhancer, a water-insoluble product is produced with a broad absorption range and a maximum at 580 nm. The reflectance of the membrane is measured with the green light emitting diode ($\lambda_{em,max} = 566$ nm).

The test was optimized with regard to sensitivity, color intensity, and assay time. The lower the concentration of TNT, the more color is developed and the lower is the reflectance. For example, a reflectance of 40% corresponds to a deep blue color.

Cross-reactivities with several nitroaromatic compounds were examined at different concentrations. Structurally related compounds with three nitro groups (such as 1,3,5-trinitrobenzene and tetryl) show high cross-reactivities. Aromatic compounds with two nitro groups have an intermediate, or no, reaction. RDX and HMX are inactive. The difference in the cross-reactivities between immunofiltration and microtiter plate occurs because the cross-reactivities are dependent on the incubation time, and in the immunofiltration test the incubation time is substantially lower.

Groundwater samples were analyzed without preparation. Soil samples were extracted with methanol, followed by filtration with a glass-fiber filter. A quantitative color response to concentrations of TNT, in the range 1–30 μg/L in water and 50–1000 μg/kg in soil, was obtained.

Magnetic particles have been used as the solid support and means of separation in an ELISA system [229]. Covalent bonding of antibodies to magnetic particles eliminates antibody dissociation that might cause imprecision. Their size allows for even dispersion throughout the reaction mixture. This places the antibody in proximity with other reactants, creating rapid reaction kinetics. In addition, they contribute the means for separating bound from free enzyme, enabling these systems to be very simple and easy to use. In step one, 100 μL of the sample to be analyzed is added, along with 250 μL enzyme conjugate, to a disposable test tube containing 500 μL antibodies attached to magnetic particles. Any TNT present in the sample competes with the labeled TNT for binding sites on the antibodies. This immunological reaction (incubation) occurs for 15 min. In step two, a magnetic field is applied to the magnetic particles with TNT and labeled TNT bound to the antibodies. All particles are pulled and held to the tube wall, while excess reagents are decanted. The particles are washed twice. In step three, the amount of enzyme-labeled TNT is determined by adding color solution, which produces a blue color after 20 min of incubation, after which the reaction is stopped by the addition of acid. The color intensity is then read at 450 nm with a portable spectrophotometer. The lowest detectable amounts in water and in soil was 70 p.p.t. and 70 p.p.b., respectively. The cross reactivity of the assay, expressed as the dose required to displace 50%, is 2.2 for 1,3,5-trinitrobezene, 30.0 for tetryl, 35.0 for 2,4-dinitrotoluene, 45.0 for 2-amino-4,6-dinitrotoluene, 98.0 for 4-amino-2,6-dinitrotoluene and >10 000 for RDX and HMX.

6.4.2.2 Evanescent wave fiber-optic biosensor

A competitive fluorescence immunoassay was performed on the surface of a fiber-optic probe for the detection of TNT [230]. When antibodies, immobilized on the fiber surface, bind the fluorescently labeled TNT analog, laser light in the evanescent wave excites the fluorophore, generating a signal. TNT, present in the sample, prevents such binding, thereby decreasing the signal.

Trinitrobenzenesulfonic acid (TNB–SO$_3$H), an analog of TNT, was labeled with a fluorophore and used as the analyte competitor. A solution containing 7.5 ng/mL sulfoindocyanine 5-ethylenediamine-labeled TNB–SO$_3$H (Cy5–EDA–TNB–SO$_3$H) was exposed to an antibody-coated optical fiber, generating a specific signal above background that corresponds to the reference (100%) signal. When TNT is present in the sample, the binding is prevented, thereby decreasing the signal. Inhibition of the reference signal is proportional to the TNT concentration in the sample.

The fiber-optic fluorometer is shown in Figure 6.10. A 650 nm diode laser is used as the excitation source. The laser light is reflected off a dichroic mirror selected for Cy5 fluorescence detection and directed into the proximal end of the fiber probe. The returning fluorescence passes trough the dichroic mirror and a bandpass filter (682 nm) onto the photodiode. A chopper modulates the laser light to enable synchronous detection with a lock-in amplifier. An electronic shutter blocked the laser light between readings to prevent photobleaching of the fluorophore.

The fiber-optic probe is prepared by removing the cladding from the last 12 cm of a 1 m long, 200 μm fused silica optical fiber, to form the sensing region. This region is tapered by etching with hydrofluoric acid, cleaned, and coated with antibodies, using a method developed previously [231]. The specific antibody used in this study was the monoclonal antibody 11B3, which was raised in TNB–SO$_3$H–ovalbumin.

Competitive immunoassays were performed to determine the concentration of several compounds, structurally similar to TNT, which were required to cause a decrease in the reference signal, similar to that achieved with 150 ng/mL (i.e. 50% inhibition). Compounds tested included degradation products of TNT, which did compete with Cy5–EDA–TNB-SO$_3$H for binding to the antibody. Concentrations of 400 ng/mL dinitrobenzene, 450 ng/mL 2,4-DNT or 2,6-DNT were required to achieve 50% inhibition of the reference signal.

Detection sensitivity of 10 μg/L of TNT in water samples was obtained. The test sample was assayed in under 4 min, while the complete cycle of the reference signal, test sample signal, and regeneration of the fiber-optic, was completed in 10 min. Less than 0.5 mL of aqueous sample was required for each assay. The ability to do multiple analyses on the same fiber probe, permits the reference signal and the competitive test signal to be run on the same fiber.

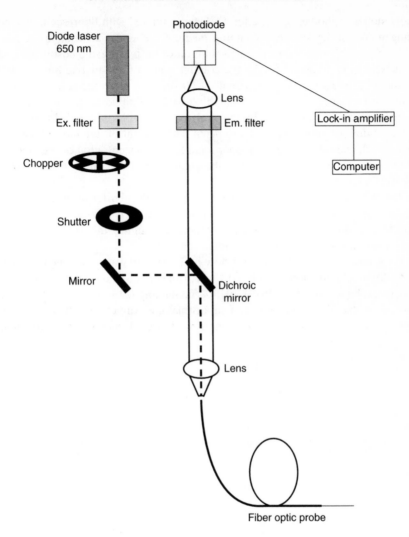

Figure 6.10 Schematic diagram of fiber-optic fluorometer. Reprinted with permission from Shriver-Lake, L. C., et al., *Anal. Chem.*, **67**, 2431 © 1995 American Chemical Society.

An improved portable fiber optic biosensor was used for on-site analysis of TNT in groundwater [232]. The biosensor was able to detect 20 µg/L TNT in less than 16 min simultaneously on four probes.

6.4.2.3 Continuous-flow immunosensor

A continuous-flow immunosensor for the detection of TNT in aqueous solutions was developed [233]. The sensor is a column containing immobilized monoclonal antibodies specific for TNT and saturated with fluorescein-labeled antigen. When the analyte containing medium (generally an aqueous sample which is spiked with a concentrated solution of buffer, organic cosolvent, and a surfactant) is pumped through the column, some of the analyte binds to the antibody, displacing the labeled analog. The effluent from the column is thus a mixture of analyte which did not bind to the antibodies and dye-labeled antigen which was displaced by the analyte. The displaced labeled antigen is detected downstream using a fluorometer. The amount of dye-labeled antigen which is displaced, is directly proportional to the concentration of the amount of analyte in the sample applied to the column, over the range 10–40 ng/mL for TNT and 20–1200 ng/mL for DNT.

The flow immunosensor (Figure 6.11) consists of a buffer reservoir, a peristaltic pump, a Rheodyne five-way valve (used as a low-pressure sample injector), a 4.6 mm inner diameter column containing the antibody-coated Sepharose, a spectrofluorometer, and an integrator.

For sample testing, a continuous-flow buffer stream was established through the column at a flow rate of either 0.4 or 1.0 mL/min. Standard flow buffer was phosphate-buffered saline (PBS), pH 7.4, containing 0.1% Triton X-100. Four solvents (ethanol, 2-propanol, dimethylformamide, dimethyl sulfoxide) were tested individually as additives in the flow buffer. Ethanol and 2-propanol,

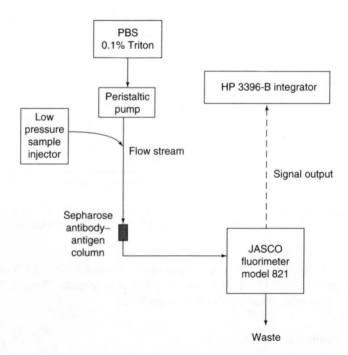

Figure 6.11 Continuous-flow immunosensor. Reprinted with permission from Whelan, J. P., et al., *Anal. Chem.*, **65**, 3561, © 1993 American Chemical Society.

when added to the flow buffer at concentrations of up to 25% (v/v), were found to increase the sensitivity by one order of magnitude. This effect was probably due to increasing both the solubility of TNT and the quantum yield of fluorescein. Columns were washed of excess labeled antigen until a steady baseline signal was established. At both flow rates, baselines were established in less than 30 min. Stock solutions of TNT and DNT were prepared in ethanol. Phenylalanine (5 µg/mL), which like TNT has a benzene ring, and either glysine or lysine (8 µg/mL) were used routinely as negative controls. Samples were diluted in the flow buffer and introduced into the flow stream using the low-pressure sample injector.

At a flow rate of 0.4 mL/min, the flow sensor gave a positive signal above background at concentrations at low as 0.25 ng of TNT.

The continuous flow immunosensor (CFI) was used for the detection of TNT and RDX in groundwater [234, 235]. The microcolumn contained 100 µL of polyacrylamide beads to which monoclonal antibodies against the explosives of interest were covalently attached. The beads averaged 60 µm in diameter and contained an azlactone group which is capable of forming a stable amide bond with any primary amine group on the antibody. The antibodies used were the 11B3 strain of monoclonal anti-TNT antibody [233] and the 50518 strain of monoclonal anti-RDX antibody. A sulfoindocyanine dye was chosen as the fluorescent dye because it has a large quantum yield (>0.28), because it is water soluble, due to the two sulfonate groups present, and because its emission maximum is far into the red (667 nm), where there is little interference from naturally-occurring fluorescent species [236]. Detection of the dye-labeled species is achieved using a fluorometer. The species in the flow cell are excited at 632 nm, while the fluorescence is recorded at 663 nm. When the analyte of interest is not present in the sample, the signal output from fluorometer vs. time, is a flat line, indicating that only background fluorescence was detected. If the explosive is present in the sample, a peak in the fluorescence vs. time trace will be observed. There is a direct correlation between the peak area and the amount of analyte present in the sample. The amount of sample required for analysis is less than 1 mL.

The limits of detection were found to be 50 and 20 p.p.b. for TNT and RDX, respectively.

HMX present in the samples did not affect the measurements of RDX.

Figure 6.12 shows a contour plot for RDX present in samples taken from wells in a field trial site which had been a munitions storage and handling depot since World War II. CFI and HPLC results were compared, and showed good agreement. The area covered was approximately 113 acres ($4.6 \times 10^5 \, m^2$).

6.4.2.4 Capillary-based displacement flow immunosensor

A capillary-based displacement flow immunosensor for the detection of TNT has been developed [237, 238]. Anti-TNT antibody is immobilized onto the silanized inner walls of a 0.55 mm inner diameter, 20 cm long,

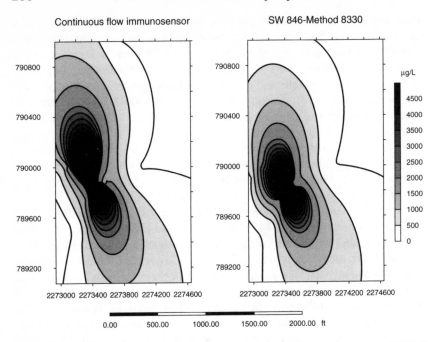

Figure 6.12 Contour plots for RDX in samples taken from wells, by CFI and HPLC (EPA Method 8330). Reprinted by permission from Bart, J. C., et al., *Environ. Sci. Technol.*, **31**, 1505, © 1997 American Chemical Society.

fused-silica capillary column, using a heterobifunctional cross-linker, followed by saturating the capillary with fluorophore-labeled antigen. The capillary was incubated with a 4% (3-mercaptopropyl)trimethoxysilane (MTS) solution (in toluene) at room temperature for 1 h. After flushing the capillary with toluene three times, a 2 mM cross-linker solution (*N*-succinimidyl 4-maleimidobutyrate (GMBS); 15 mg in 27 mL of ethanol) was introduced in the capillary and incubated at room temperature for 1 h. The capillary was then rinsed with deionized water followed by the addition of 1 mg/mL solution of anti-TNT antibody 11B3. After 1 h, the capillary was rinsed with deionized water three times. Finally, a 30 μM Cy5-TNB solution was introduced into the capillary and incubated in a refrigerator at 4°C for a minimum of 12 h.

Before an assay, a small fraction of the sample solution (TNT-spiked buffer solution or soil extract) was pumped into the 16 μL flow cell of the fluorometer, to confirm that the sample does not contain any fluorescent contaminants, which could interfere with the TNT immunoassay. The excitation wavelength used was 632 nm and the fluorescence emission was monitored at 662 nm, which are the excitation and emission maxima, respectively, for the Cy5 dye.

The next step was to pump the flow buffer (a mixture of sodium monophosphate, ethanol and Tween) through the capillary column and monitor the fluorescence of the Cy5-TNB solution. Initially, there was a constantly changing slope for the rate of change of in fluorescence intensity as a function of time, indicating the washing of the unbound and nonspecifically adsorbed Cy5-TNB from the walls of the column. After 20–30 min, the background fluorescence stabilized and remained constant. Now, 100 µL injections of TNT-containing flow buffer samples were made in the loop injector. TNT concentrations ranged from 1 pg/mL to 1000 ng/mL. The resulting peaks due to the displacement of the fluorescent antigen were recorded, and the integrated area under the peaks used for quantitation. The optimal buffer flow rate was in the range 100–250 µL/min. TNT was extracted from soil by shaking the soil in acetone, followed by filtering through a glass microfiber filter: 200 µL of the extract was dried and resuspended in the flow buffer.

The limit of detection was 440 amol (1×10^{-18} mol) (100 µL injection of 1 pg/mL TNT solution). The entire assay could be done in less than 3 min. The relative standard deviation (RSD) of the area under the peaks, resulting from ten injections of 5 ng/mL into the same column was 4%. The RSD of displaced area under the peaks upon injecting 100 µL of 5 ng/mL TNT into five capillary columns was 5%.

Relative to TNT (100%), the cross-reactivities of related compounds were as follows: TNB (85%), 4-A-2,6-DNT (21%), DNT (23%), nitrotoluene (15%), nitrobenzene (13%), RDX (18%), HMX (12%), phenylanaline (3%), and glycine (2%). Comparison of contaminated soil samples, tested with the capillary immunosensor and with HPLC, showed good agreement.

6.5 ELECTROCHEMICAL DETECTION METHODS

6.5.1 Properties of amperometric gas sensors

Amperometry is an electroanalytical technique that is widely used to identify and quantify electroactive species [239, 240]. In amperometric gas sensors, measurements are made by recording the current in the electrochemical cell between working (or sensing) and counter (or auxiliary) electrodes at a certain potential. These devices are distinguished from potentiometric sensors in which the potential at (near) zero current flow, is the measured signal, or conductometric sensors, in which the measured signal are changes in impedance. Amperometric sensors produce a current when exposed to a vapor containing an electroactive analyte reacting within the cell, either producing or consuming electrons, in which case the analyte is electro-oxidized or electro-reduced at the electrode. This process can be accelerated by a catalyst, such as a Pt electrode or an enzyme electrode, or be a sacrificial process where the electrode material is consumed. By choosing a suitable geometry for gas exposure and electrolyte confinement, appropriate electrodes and the electrochemical method (e.g. fixed potential, pulsed potential), the sensor can be made quite sensitive for a particular analyte and insensitive to others.

In most amperometric gas sensors, the working electrode is not directly exposed to the vapor stream; the analyte must diffuse through a porous membrane or frit. This interface serves also as a barrier to prevent leakage of electrolyte from the interior of the sensor. Following dissolution into the electrolyte, the analyte migrates to the working electrode, where it is adsorbed on its surface and available for electrochemical reaction.

As an example a CO sensor will be used for demonstration. The fundamental process in the electrochemical reaction is the transfer of electrons from CO to the electrode surface, coupled with a chemical reaction in which an additional oxygen-carbon bond is formed as follows:

$$CO + H_2O \longrightarrow CO_2 + 2H^+ + 2e^- \tag{6.4}$$

A second reaction, reduction of oxygen to water, occurs at the counter electrode:

$$\tfrac{1}{2} O_2 + 2H^+ + 2e^- \longrightarrow H_2O \tag{6.5}$$

This counter reaction is fundamentally necessary to maintain charge neutrality in the electrochemical cell.

The electrochemical reaction products will desorb from the working electrode and diffuse away. This process cleans the working electrode area and returns this area to its initial stage.

A potentiostat is used with a three-electrode sensor to provide a fixed or controlled potential for the working electrode, relative to a reference electrode, which is placed in the same electrolyte. The potentiostat controls the electrochemical cell and converts the sensor's current signal to a voltage signal. In a two-electrode system, the auxiliary electrode serves as both counter and reference electrodes. The three-electrode configuration allows for precise operation. The reference electrode is used to maintain the working electrode at a known potential; it must therefore be physically and chemically stable. The working electrode (i.e. a noble metal gas-diffusing electrode) must be able to keep the electrolyte in the cell, but must be porous enough to allow the gas phase to diffuse to the interface of electrode and electrolyte. The circuit must be designed so that virtually no current passes through the reference electrode. The counter electrode completes the electrochemical cell by performing the half cell reaction. The nature of this reaction must be opposite to the working electrode reaction in order to minimize net chemical changes in the sensor.

6.5.2 Amperometric gas sensor for the detection of thermal decomposition products of TNT

Thermal decomposition of explosives over heated noble metal surfaces generates characteristic products that can be detected by amperometric gas sensors. This effect was used to develop a method for *in situ* detection of TNT in contaminated soils [241]. As gas sensors are not sensitive enough to measure directly TNT vapor, a system was developed in which characteristic catalytic degradation products of TNT undergoing pyrolysis are detected.

The distribution of thermal products depends primarily on the temperature and nature of the catalytic surface, which can be controlled, and on the compound undergoing pyrolysis. The characteristic vapors generated by pyrolysis of TNT and other explosives [242] contain NO and other nitrogen–oxygen compounds, in addition to carbon dioxide, carbon monoxide, and water.

The system which was developed could be installed inside a cone penetrometer, in order to measure subsurface contamination by explosives in soils. The principle of detection consists of three steps:

1. vaporization of the explosive,
2. thermal decomposition of the explosive vapor,
3. measurement of nitric oxide (NO) on miniature amperometric sensors.

For suitable amperometric operation, the analyte must be electrochemically active and must come into physical contact with the working electrode of the sensor.

A laboratory probe was built for the thermal decomposition of soil-bound TNT and measurement of characteristic decomposition products (Figure 6.13). About 500 g of the soil tested was placed in a PVC vessel containing a heater element consisting of a platinum wire wrapped around a 0.5 in diameter alumina tube. Alternative catalytic materials, such as rhodium and nichrome were evaluated, but their performance was inferior to that of platinum. A pneumatic system collected and transported the vapor generated by thermal decomposition of the heater, to the sensor. The air flow through the ports and

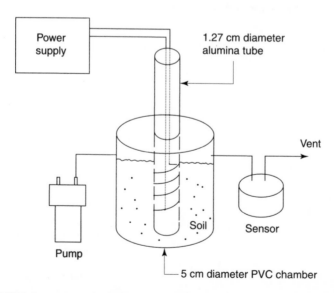

Figure 6.13 Laboratory probe for thermal decomposition of soil-bound TNT. Reprinted from Buttner, W. J., et al., *Anal. Chim. Acta*, **341**, 63, © 1997, with permission from Elsevier Science.

over the sensor was maintained at a rate of 200 cm^3/min for the duration of the measurement. Amperometric sensors with high sensitivity gold sensing electrodes were used as detectors. The sensor was controlled by a conventional potentiostat circuit design, modified for low noise and low power operation. Before exposure to vapors, the sensor was biased to the desired potential and allowed to stabilize for at least 15 min. All measurements were made at constant sensor potential. Heater temperature was set at 900°C, by means of a 30 V, 10 A power supply. For field tests, a penetrometer system probe was designed to accommodate both an electrochemical sensor for the *in situ* measurement of explosive contamination and geophysical sensors for determining soil mechanical properties [243]. The probe, shown in Figure 6.14 [244], was made from a 2 in outer diameter, 1.5 in inner diameter stainless steel tube. Thermal decomposition of explosives was obtained through a 20 cm long platinum heater wire wound around a high-temperature ceramic, and operated at 900°C. A sacrificial sleeve protects the gas flow ports and the pyrolyzer unit, as the probe is advanced into the soil; it is left behind

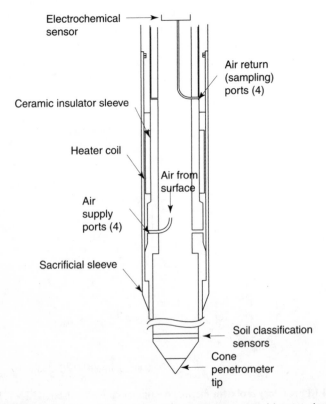

Figure 6.14 Penetrometer explosives probe. Reproduced with permission from Wormhoudt, J., et al., *Appl. Optics*, **35**, 3992, © 1996 Optical Society of America.

as the probe is withdrawn. The withdrawal of the probe is halted at desired depths, and the pyrolyzer is activated for periods of the order of 1 min. Air is continuously pumped from the surface through the air supply ports of the probe, collected through the air return or vapor sampling ports, and directed over the electrochemical sensor located in the probe. The analog signal from the electrochemical sensor, which is proportional to the NO concentration, is transmitted to the data acquisition system located in the cone penetrometer truck. An amperometric sensor with gold electrodes is used, which detects selectively nitrogen oxides (and does not detect carbon monoxide and carbon dioxide). TNT-containing soils produced electrochemically oxidizable vapors which were proportional to the concentration of TNT in the soils, while TNT-free soils produced negligible responses. Fixed potential operation does not differentiate between NO, NO_2, or any other oxygen and nitrogen-containing compounds. Once identified, the sensor system could be optimized around the actual electroactive compound, in order to maximize sensitivity, selectivity, and response time. NO was found to be the primary electroactive product formed from thermal catalysis of TNT. Accordingly, the bias to the gold sensor was adjusted from -200 to $+400$ mV (vs. the Pt–air reference electrode in the sensor).

In a typical measurement, the sensor baseline was measured for 30 s, followed by activation of the heater to 900°C. Only a brief heater activation time was necessary to produce a detectable response in the presence of TNT. Activation times longer than 30 s depleted TNT from around the probe, resulting in a decaying signal, even during the heater activation.

The sensor response was reproducible for a give concentration of TNT, with a typical precision of $\pm 10\%$. The total analysis time was less than 4 min. The sensor response is proportional to the soil concentration of TNT over the range 1–500 p.p.m.-wt. The sensor response factor was 10 mV/p.p.m. throughout this range.

It was found that the analyte had to be in contact with the active catalytic surface to generate a sensor response. As a result the detector measures only TNT which is in contact with the pyrolyzer and will not measure contamination over a large volume.

An attempt to use a rhodium-based heater instead of platinum showed the ability of rhodium to physically trap TNT, but did not result in sensor peaks as intense as with platinum. It was concluded that the generation of decomposition products was not simply a thermal degradation process, but that the catalytic activity of the heater surface controls to some extent the distribution of the product compounds.

The electrochemical TNT detector was field tested at a site of a former munitions production facility that had been in operation for 50 years. Over 14 locations were analyzed. At each location, measurements were made from 15 ft below surface to the surface, providing a rapid depth profiling of TNT. Differences between results obtained with the electrochemical detector and alternate methods, including HPLC, were attributed to heterogeneity of TNT in soil samples.

6.5.3 Cyclic voltammetry sensor for the detection of TNT in water and soil

A potentiodynamic detector based on cyclic voltammetry for the detection of TNT in water and soil was developed [245]. The electrochemical sensor is based on a triple electrode array. The tip of a 25 µm gold wire served as working electrode. The reference electrode was a saturated calomel electrode and the counter electrode a gold foil. The potential of the working electrode is varied periodically. The base electrolyte was 5 M H_2SO_4.

TNT is reduced at 0.2 V. The reduction peak is not available as a sensor signal. In the anodic scan an oxidation peak at approximately 0.7 V is observed. Although the oxidation current is smaller than the reduction current, the oxidation current is proportional to the TNT concentration, and serves as sensor signal. TNT as well as DNT, could be detected in soil at concentrations of at least 1 mg/kg.

A complete measurement, including soil sampling, took 5 min.

In addition to TNT, additional pollutants are found in the soils, such as aminoaromatic compounds. Potentiodynamic experiments were carried out with 2,4-DNT, 2,6-DNT, 1,3-dinitrobenzene, 2-amino-5-nitrobenzene, and 2,4-diaminotoluene in solution. Cyclic voltammograms were recorded and the potentials of the current peaks determined. The potential differences between the oxidation peaks of 2,4-DNT, 2,6-DNT, and 1,3-DNB and the TNT oxidation were too small for a quantitative differentiation. Therefore, nitroaromatic compounds were determined as a total parameter. On the other hand, the potential differences between the oxidation peaks of TNT and 2-amino-5-nitrobenzene and 2,4-diaminotoluene, allow the distinction between amino- and nitroaromatic compounds.

Another electrochemical sensor, for remote continuous monitoring of TNT, was based on square-wave voltammetry at a submersible carbon-fiber electrode assembly connected to a 50 ft-long shielded cable [246]. This design uses also a three-electrode assembly. The carbon fibers were pretreated by dipping into 6 M nitric acid for 10 s, then rinsed with distilled water. Another wash with acetone was followed by rinsing the fibers with distilled water. The reference electrode was a silver–silver chloride electrode and the counter electrode was a platinum wire.

Experiments were conducted in the laboratory, while placing the sample 50 ft from the voltammetric analyzer. The electrode assembly was fully submersed in the 40 mL river water sample, placed in a 60 mL beaker. Square-wave voltammograms were recorded over the 0.0–(−0.8) V range, using a 30 Hz frequency, 10 mV amplitude, and a 2 mV step height. These correspond to a 12 s scanning time, that is 300 measurements per hour.

Unlike laboratory-based measurements, where the solution conditions can be adjusted for optimal performance, submersible probes rely on the use of natural conditions (i.e. untreated samples).

The peak currents were found to be directly proportional to the TNT concentration up to 7 mg/L and increase more slowly thereafter. Detection limits were estimated to be around 200 and 30 μg/L for river and drinking water samples, respectively, with a signal-to-noise ratio of 3.

While the nitro group is rare in nature, nitro-containing pollutants, including related explosives, pose a potential interference. The presence of RDX, nitroguanidine, nitrophenol, nitrobenzene, and 3,5-dinitroaniline, at 10 mg/L, resulted in 4, 3, 7, 32, and 5% suppressions, respectively, of the 5 mg/L TNT response. Most of these compounds also displayed additional peaks at higher potentials.

Screen-printed carbon electrodes were developed as disposable voltammetric sensors for TNT [247]. The method is based on placing the thick-film carbon sensor in the water containing TNT and using a fast (<1 s) and sensitive square-wave voltammetric scan. A single voltammetric peak at −0.45 V was obtained, corresponding to the formation of the hydroxylamine moiety. Optimal parameters included an amplitude of 25 mV, a frequency of 50 Hz, and a staircase step height of 4 mV. The sensor did not show a dependence on ionic strength for phosphate buffer concentrations ranging from 10 to 200 mM, but displayed a large decrease in the current response and large potential shift without any electrolyte. However, environmental water samples usually contain sufficient natural electrolyte for direct monitoring of TNT.

The response for TNT was not affected by the presence of similar nitroaromatic compounds, such as nitrophenol and nitrobenzene, nor by the presence of nitrite or nitrate ions. A detection limit of 100 p.p.b. was obtained when using computerized background-subtraction. Figure 6.15 shows background-subtracted square-wave voltammograms for untreated tap

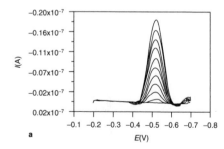 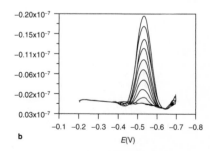

Figure 6.15 Background-subtracted square-wave voltammograms for: a, untreated tap, and b, river water samples, containing increasing levels of TNT in 200 ppb steps. Reprinted from Wang, J., et al., *Talanta*, **46**, 1405, © 1998, with permission from Elsevier Science.

(a) and river (b) water samples, containing increasing levels of TNT in 200 p.p.b. steps.

6.6 LASER DETECTION METHODS

A system based on tunable infrared laser differential absorption spectroscopy (TILDAS) with multipass, long-path absorption cells, was investigated for the detection of pyrolysis products of explosives in soils [244]. The laser is used as a sensor for the detection of NO, which is a major thermal decomposition product of TNT.

Figure 6.16 shows the components of an initial laboratory experiment which involved heating of a contaminated soil sample, and sampling of the resulting gaseous products into a multipass cell, where their laser absorption is measured as a function of time. An example of pyrolysis product concentration in TNT, as measured with the experimental set-up is shown in Figure 6.17. The two traces are cell concentrations of NO and NO_2, derived from specific lines in

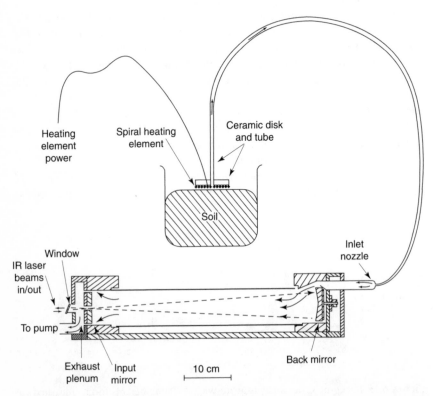

Figure 6.16 Schematic diagram of laboratory experiment for laser detection of pyrolysis products of explosives-contaminated soil. Reproduced with permission from Wormhoudt, J., et al., *Appl. Optics*, **35**, 3992, © 1996 Optical Society of America.

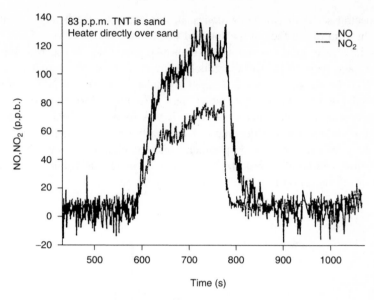

Figure 6.17 Pyrolysis product concentration of NO and NO₂ of TNT in sand. Reproduced with permission from Wormhoudt, J., et al., *Appl. Optics*, **35**, 3992, © 1996 Optical Society of America.

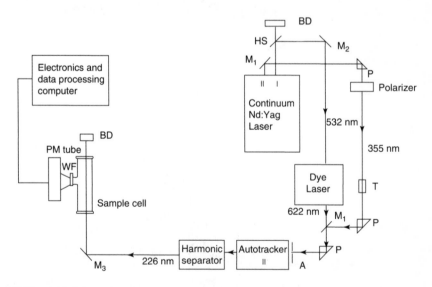

Figure 6.18 Schematic diagram of photofragmentation–laser-induced fluorescence (PF–LIF) experimental setup. PM = photomultiplier, BD = beam dump, WF = window and filter, M₃ = dichroic mirror (226 nm), HS = harmonic separator, M₁ = dichroic mirror (335 nm), M₂ = dichroic mirror (532 nm), P = prism, T = telescope, A = aperture. Reproduced with permission from Wu, D., et al., *Appl. Optics*, **35**, 3998, © 1996 Optical Society of America.

the spectral scans of two laser diodes. In this example, the heater was about 0.25 cm from the surface of the soil and was heated to 900°C for 3.5 min. The soil sample was sand, contaminated with 83 p.p.m. TNT. It can be seen that the NO and NO_2 traces show parallel behavior, but the NO signal is larger. Low NO_2 levels are attributed to rapid secondary reactions of NO_2 produced in initial bond-breaking steps, rather than to reduced formation of NO_2 in the early stages of TNT decomposition. It was found that TNT is vaporized from the soil matrix, but decomposes thermally only at higher temperatures near the pyrolyzer filament. Experiments and calculations showed that useful absorption path lengths could be obtained, even under the volume constraints imposed by inclusion in a cone penetrometer probe.

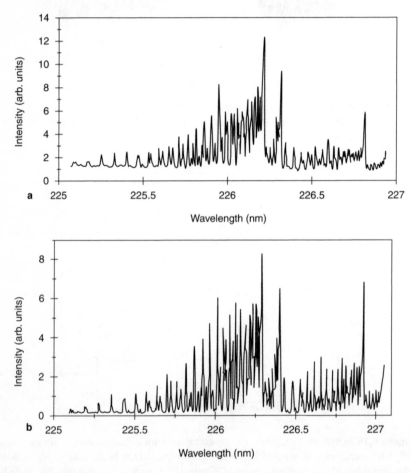

Figure 6.19 PF–LIF spectra from: a, a pure TNT sample and, b, from 100 p.p.m. TNT in soil, both recorded at 100°C. Reproduced with permission from Wu, D., et al., *Appl. Optics*, **35**, 3998 © 1996 Optical Society of America.

Photofragmentation (PF) and subsequent nitric oxide (NO) laser-induced fluorescence (LIF) has been suggested for the detection of TNT in soil [248, 249]. The technique is based on molecular PF and fluorescence spectrometry. TNT itself has a low luminescence quantum yield, and is therefore difficult to detect by fluorescence.

By the use of laser radiation, TNT can be photodissociated to NO_2 (X^2A_1), which absorbs one photon of 226 nm radiation and undergoes a transition to the 2^2B_2 state. The 2^2B_2 state is rapidly predissociated to produce NO ($X^2\Pi$, $v'' = 0$) and O. The ground-state NO then absorbs another 226 nm photon to undergo a resonant transition, producing NO fluorescence by LIF. As the NO fluorescence intensity is proportional to the TNT concentration in the sample, the TNT concentration can be evaluated from the NO fluorescence signal intensity. A schematic diagram of the PF–LIF experimental setup is shown in Figure 6.18. A frequency-doubled Nd–YAG laser was used to pump a dye laser. The 226 nm UV laser beam was generated by frequency mixing. The 622 nm dye laser output was mixed with a frequency-tripled Nd–YAG laser beam (355 nm) to obtain the desired UV laser wavelength. The UV laser beam generated, was separated from the laser beams through a four-prism harmonic separator. The sample cell was made of stainless steel with two UV-grade fused silica viewports. The PF–LIF signal was detected with a photomultiplier tube. The PF–LIF spectrum was recorded by scanning the dye laser.

Figure 6.19 shows the PF–LIF spectra from a pure TNT sample and from 100 p.p.m. TNT in soil, both recorded at 100°C. The PF–LIF spectrum of pure TNT is similar to that of NO gas with the same experimental setup. Results showed that by heating the TNT sample, the signal intensity increased. But heating the TNT sample above 70°C, produced physical and chemical changes in the sample. The limit of detection of TNT in soil was found to be 40 p.p.b. The presence of water was found to quench the fluorescence in the cell and give inconsistent PF–LIF signals when added to a TNT-containing soil sample.

7

Detection of landmines

7.1 INTRODUCTION

With the invention of durable, modern fused munitions about 150 years ago, a long-term munition safety hazard was created at battlefields and training areas worldwide. That hazard remains long after the soldiers have left. In Europe, as a result of the two world wars, accidents involving forgotten munitions are not uncommon. In France, 630 explosive ordnance disposal specialists have been killed since 1946. During 1991, 36 French farmers died when their machines struck unexploded shells.

Worldwide, a subset of munitions, landmines, is a particular problem. According to the United Nations, 120 million unexploded landmines are buried in 70 countries, most of them unmarked. About 24 000 civilians are killed or maimed by those devices every year, which means a victim every 20 min. The UN estimates that 80 000 were cleared in 1995; during the same period, 2.5 million more were installed.

The mine menace is worst in the following countries.

- Afghanistan: 10 million mines
- Angola: 15 million mines
- Bosnia and Herzegovina: 3–6 million mines
- Cambodia: 10 million mines
- Croatia: 3 million mines
- Egypt: 23 million mines
- Iran: 16 million mines
- Iraq: 10 million mines
- Mozambique: 2 million mines

The price of a landmine is about $3–15, while the cost of clearing it can be as much as $1000 [250]. Postconflict clearing of mines is very slow. The UN estimates that it will take $33 billion and 1100 years to clear all the mined areas in the world with current technology.

7.2 CLASSIFICATION OF MINES

The concept of the landmine is very simple: a sensor detects a target, which activates a sensitive detonator, which in turn sets off a larger quantity of a high explosive. All the components are encased in metal, plastic, or wood. The fuse, which includes the sensor and the activating mechanism, can be configured in many ways. Fuse technology has evolved from simple mechanical pressure sensors to electronic sensors that can analyze a wide range of signals from their targets, such as acoustic, seismic, magnetic, or thermal. Some mines, known as 'smart' or 'sensory' devices, incorporate microprocessors and are able to distinguish between animals and human beings, while others count the number of people who have passed before setting themselves off.

The most used explosive for the main charge is TNT. Other explosives used are RDX, Composition B, tetryl, and C-4. In addition, mines have a booster charge to enhance the power released by the detonator.

Two categories of landmines are commonly used:

- Large antitank mines, designed to be triggered by vehicles. They contain 2–10 kg of explosive and are activated by pressures of 100–300 kg. Some of these mines are modified to be detonated by only 7 kg, which makes them very powerful antipersonnel mines.
- Anti-personnel landmines are smaller, 7–15 cm diameter, with 10–250 g of explosive, and detonate under pressures of 0.5–50 kg. They can be subdivided into three groups.
  — Blast mines, which burst on contact, attacking a single individual. They rely on the energy released by an explosive charge to harm their target. They are usually buried by hand or placed on the ground. Most modern antipersonnel blast mines have a plastic watertight casing; only the detonator, springs and strikers are metal, making them difficult to detect.
  — Bounding mines, which leap to a height of 2 m before blowing up with a destructive range of 30 m.
  — Directional fragmentation mines, which erupt in a spray of shrapnel, propelling deadly fragments in an arc that reaches as far as 200 m. These mines are usually activated by tripwires, which may be placed a few centimeteres above the ground. An example is shown schematically in Figure 7.1 [251].

Mines may be planted manually or they may be delivered from the air, by mortar, rockets or artillery. Mines may also be dispensed from containers carried under helicopters or from underwing dispensers on low-flying fixed-wing aircraft [252]. Some devices are used to hurl mines mechanically over large areas and considerable distances. One vehicle-mounted mine-scattering system can dispense up to 1750 antipersonnel mines per minute, while a helicopter-mounted system is designed to drop 2080 antipersonnel mines in 3–16 minutes [253].

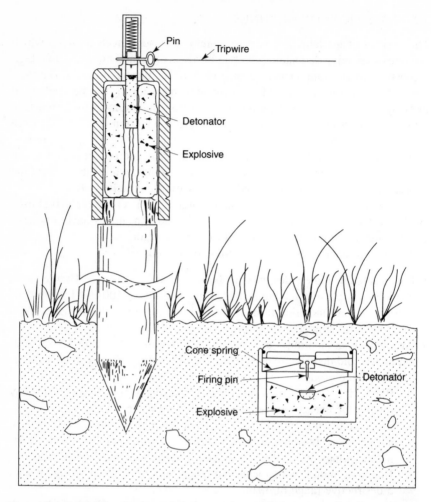

Figure 7.1 Schematic diagram of fragmentation mine activated by tripwire. Reproduced from Trevelyan, J., University of Western Australia, Department of Mechanical and Materials Engineering, Nedlands 6907, Western Australia, 1997, with permission.

The inherent difficulty of mine clearance is complicated by the great variety of mines in use: more than 700 types are known.

It is beyond the scope of this book to give an exhaustive description of all the mines known. Instead, some typical examples are presented. Table 7.1 shows examples of landmines in use [254].

Figure 7.2 shows schematically 12 types of common antipersonnel landmines [252]. Their description is presented in Tables 7.2–7.13.

The most prevalent of the conventional landmines are the blast mines. They rely upon the energy released by the explosive charge to harm their target, and are normally buried by hand or placed on the ground. The injuries they produce

Table 7.1 Examples of landmines in use. Reprinted from Tsipis, K., *Report on the Landmine Brainstorming Workshop*, Report No. 27, MIT, Program in Science and Technology for International Security, November 1996, with permission

Country	Name	Main charge (g)	Booster (g)	Type	Construction
Italy	Valmara 69	Comp B (576)	RDX (13)	Bounding	Plastic cylinder
	BM-85	Comp B (450)		Bounding	Plastic
	VS-50	RDX (43)		Blast	Plastic disc
Russia	PMN	TNT (240)	Tetryl (9)	Blast	Plastic disc
	PFM-1	liquid (35)		Blast, scatter	Green parrot
	PMD-6M	TNT (200)		Blast	Wooden box
	POMZ-2	TNT (75)		Fragment	Wooden stake
	MON-100	TNT (1890)		Directional	Metal disc
	OZM-72	TNT (700)	Tetryl (23)	Bounding	Steel cylinder
USA	M14	Tetryl (28)		Blast	Plastic disc
	M18A1	C-4 (682)		Directional	Plastic
	M16	TNT (521)	Tetryl	Bounding	Metal cylinder
China	Type 69	TNT (105)		Bounding	Iron cylinder
	Type 72	TNT or TNT/ RDX(50/50) (51)	RDX (24)	Blast	Plastic disc
Czech Republic	PP Mi-Sr	TNT (325)		Bounding	Metal cylinder
	PP Mi-Sb	TNT (75)		Fragment	Concrete stake
	PP Mi-Ba	TNT (200)		Blast	Plastic cylinder
	PP Mi-D	TNT (200)		Blast	Wooden box
Germany	PPM-2	TNT (110)		Blast	Plastic disc
Yugoslavia/ Bosnia	PROM-1	TNT (425)		Bounding	Steel bottle
	PMA-1	TNT (200)		Blast	Wooden box
	PMA-2	TNT (100)	RDX (14)	Blast	Plastic cylinder
Austria	APM-1	Comp B (360)		Directional	Tripod mount
France	M-59	TNT (57)	Tetryl (17)	Blast	
	M-61	TNT (57)		Blast	Plastic stake
	M-63	Tetryl (30)		Blast	Plastic stake

result primarily from the explosion, but secondary fragmentation injuries are possible as the mine casing or surrounding dirt or gravel is blasted at the victim.

Some mines additional to those shown in Figure 7.2 are the following [252]:

- MON-50 is the Russian version of the American M-18, antipersonnel directional fragmentation mine. It has a curved shape and can be mounted against

Figure 7.2 Common antipersonnel landmines. Reproduced from Publication No. 10225, US Department of State, Bureau of Political-Military Affairs, 1994.

a round surface, such as a tree, or can be placed on a small stand-alone stake. The mine is filled with pellets or projectiles in front of the explosive charge, which are shot out by the blast at a high velocity and can kill targets up to 50 m away. They can be activated by a simple remote-control switch.

- PFM-1 is a Russian scatterable pressure-sensitive blast mine. It has a butter-fly-like shape, which attracts children who think it is a toy. It has been produced in various shades of brown, green, and white. The PFM-1S version of this mine is one of the rare designs which include a self-destruct mechanism. It explodes 24 h after deployment.
- OZM-4 is a metallic bounding fragmentation mine, which is designed to kill the person who sets it off, and to injure anybody nearby by propelling fragments. The cylindrical mine body is initially located in a short pot or barrel assembly; activation detonates a small explosive charge, which

Table 7.2 Type 69 antipersonnel mine. Reproduced from *Hidden Killers 1994: The Global Landmine Crisis*, P.F. Schultz III, Editor, Publication No. 10225, US Department of State, Bureau of Political-Military Affairs, 1994

General	
Mine type	Antipersonnel
Lethal effect	Fragmentation
Manufacturing country	China
Characteristics	
Shape	Cylindrical
Height (mm) (w/o fuse)	60
Height (mm) (w/fuse)	114
Diameter (mm)	168
Total weight (kg)	1.35
Mine case	
Material	Cast iron
Number of fuse wells	1 (plus a detonator well)
Color	Olive drab
Mine fuse	
Model	Type 69
Type	Pressure or tripwire
Actuation force:	
Pressure (kg)	7–20
Tension (kg)	1.5–4
Explosive components	
Main explosive type	TNT (50/50)
Main explosive weight (kg)	0.105
Mine performance	
Effective range (m)	11
Detectability	Readily detectable due to metal case
Antidisturbance	Possible (however, no secondary fuse-well or AD features)

projects the mine body upwards. As it 'bounds' in the air, the mine is activated by an anchor cable (secured to the barrel assembly which remains on the ground) which pulls a pin from the fuse on the mine's body. The resulting blast scatters fragments over a much wider radius and area than would be possible with a surface or buried mine of similar size.

- BPD-SB-33 is a scatterable antipersonnel mine. Its irregular shape and small size (about 9 cm diameter) make is particularly hard to locate. A hydraulic antishock device ensures that it cannot be detonated by explosions or artificial pressure. It is also very light, and can thus be carried and deployed in extremely large numbers by helicopters.

Table 7.3 Type 72 antipersonnel mine. Reproduced from *Hidden Killers 1994: The Global Landmine Crisis*, P. F. Schultz III, Editor, Publication No. 10225, US Department of State, Bureau of Political-Military Affairs, 1994

General	
Mine type	Antipersonnel
Lethal effect	Blast
Manufacturing country	China, South Africa
Characteristics	
Shape	Cylindrical
Height (mm)	38.5
Diameter (mm)	78.5
Total weight (kg)	0.125–0.15
Mine case	
Material	Plastic body with rubber cover
Number of fuse wells	1 (detonator well)
Color	Green body, light green cover
Mine fuse	
Model	Integral
Type	Pressure
Actuation force (kg)	5–7
Explosive components	
Main explosive type	TNT or TNT/RDX (50/50)
Main explosive weight (kg)	0.075 or 0.1
Mine performance	
Effective range (m)	Limited (typical with contact-fused, blast-effect mines)
Detectability	Very difficult with hand-held metallic detectors (limited metallic content)
Antidisturbance	Possible (especially if the 'look-a-like' versions, Type 72B and Type 72C are employed)

7.3 MINE DETECTION TECHNIQUES

7.3.1 Anomaly detectors

Anomaly detectors detect objects which are not expected in their natural environments.

7.3.1.1 Manual mine clearance (prodding)

Prodding the ground by hand is presently the only way that guarantees an exhaustive detection of any landmine. Well trained deminers poke the ground

Table 7.4 M14 antipersonnel mine. Reproduced from *Hidden Killers 1994: The Global Landmine Crisis*, P. F. Schultz III, Editor, Publication No. 10225, US Department of State, Bureau of Political-Military Affairs, 1994

General

Mine type	Antipersonnel
Lethal effect	Blast
Manufacturing country	United States, India

Characteristics

Shape	Cylindrical
Height (mm)	40
Diameter (mm)	56
Total weight (kg)	0.158

Mine case

Material	Plastic
Number of fuse wells	1
Color	Olive drab

Mine fuse

Model	Integral
Type	Pressure
Actuation force (kg)	9–16

Explosive components

Main explosive type	Tetryl
Main explosive weight (kg)	0.029

Mine performance

Effective range (m)	Limited (typical with contact-fused, blast-effect mines)
Detectability	Difficult with hand-held metallic detectors (metallic content limited to striker pin)
Antidisturbance	Possible (however, no secondary fuse well, or AD features)

with a thin steel spike every 2 cm at a shallow angle of about 30°. The resistance of the probe and the reaction of the surface define where to dig the ground around and carefully remove the mine. This technique is effective in soft ground, but less so in rocky terrain. There is also a danger that the mine might have been displaced so that its pressure-sensitive lid is facing the prod. One man can clear 20–50 m^2 per day.

7.3.1.2 Automated mine detection

An automated mine detection system, using a sensor that resembles the operation performed by manual prodding, has been designed [255]. The sensor

Table 7.5 M16A1 antipersonnel mine. Reproduced from *Hidden Killers 1994: The Global Landmine Crisis*, P. F. Schultz III, Editor, Publication No. 10225, US Department of State, Bureau of Political-Military Affairs, 1994

General	
Mine type	Antipersonnel
Lethal effect	Bounding fragmentation
Manufacturing country	United States
Characteristics	
Shape	Cylindrical
Height (mm)	203
Diameter (mm)	103
Total weight (kg)	3.57
Mine case	
Material	Steel
Number of fuse wells	1
Color	Olive drab, green
Mine fuse	
Model	M605
Type	Trip wire, pressure
Actuation force (kg)	2, 5
Explosive components	
Main explosive type	TNT
Main explosive weight (kg)	0.513
Mine performance	
Effective range (m)	27
Detectability	Readily detectable due to metallic case and fragmentations
Antidisturbance	Possible (however, no secondary fuse well, or AD features)

consists of one or more independently actuated linear modules equipped with a 30 cm long needle attached to the end of the module. Force, position and speed feedback measurements, are used to determine the depth of a buried object. Position and speed are measured by an incremental encoder. The force sensor is attached to the base of the probe to measure the ground penetration force. The depth of the buried object is determined by measuring the distance from a reference position to the point where the probe makes contact with an object. It is necessary to probe the soil at least at every 5 cm to be sure to detect every mine in the area.

A detection algorithm uses the derivative of the force function, calculated as the difference between the force measured at consecutive points. The force

Table 7.6 M18A1 antipersonnel mine. Reproduced from *Hidden Killers 1994: The Global Landmine Crisis*, P. F. Schultz III, Editor, Publication No. 10225, US Department of State, Bureau of Political-Military Affairs, 1994

General	
Mine type	Antipersonnel
Lethal effect	Directed fragmentation
Manufacturing country	United States, Chile, South Korea
Characteristics	
Shape	Rectangular
Length (mm)	216
Width (mm)	35
Height (mm)	83
Total weight (kg)	1.58
Mine case	
Material	Fiberglass reinforced plastic
Number of fuse wells	2
Color	Olive drab
Mine fuse	
Model	M57 firing device
Type	Command detonation
Explosive components	
Main explosive type	C-4
Main explosive weight (kg)	0.682
Mine performance	
Effective range (m)	50
Detectability	Visually detectable
Antidisturbance	Possible (however, no secondary fuse well, or AD features)

derivative is proportional to the soil resistance during penetration, and to the object hardness when the object is touched.

A microcontroller controls the speed of a d.c. motor drive, that actuates the sensing device. Force measurement is performed synchronously with the module position, once every millimeter of penetration.

The system has been mounted on a remote controlled robot.

7.3.1.3 Metal detectors

Most metal mine-detectors are portable and are operated by one man. They consist of a search coil, generating a magnetic field, and located on the end of an extending telescopic pole. The magnetic field may be disturbed by a

Table 7.7 Valmara 69 antipersonnel mine. Reproduced from *Hidden Killers 1994: The Global Landmine Crisis*, P. F. Schultz III, Editor, Publication No. 10225, US Department of State, Bureau of Political-Military Affairs, 1994

General	
Mine type	Antipersonnel
Lethal effect	Bounding fragmentation
Manufacturing country	Italy
Characteristics	
Shape	Cylindrical
Height (mm)	205 (with fuse)
Diameter (mm)	130
Total weight (kg)	3.3
Mine case	
Material	Plastic
Number of fuse wells	2
Color	Green, sand brown
Mine fuse	
Model	unknown
Type	Trip wire, pressure
Actuation force (kg)	6, 10
Explosive components	
Main explosive type	Comp B
Main explosive weight (kg)	0.597
Mine performance	
Effective range (m)	27
Detectability	Readily detectable due to metallic fragmentations
Antidisturbance	Probable (there is a secondary fuse well, on the bottom of the mine for boobytrap purposes)

metal object. The consequence is a higher power consumption or a change in the magnetic field induced into another coil. The information acquired by the induction coil sensor can be converted into an audio signal or can be used for imaging purposes. By moving the sensor and displaying data using different colors for different response amplitudes, a map of the metal content in the soil can be constructed.

Larger and more sensitive search coils and related systems may be carried in arrays on the front of vehicles, usually with a control device, which will stop the vehicle when a suspicious object is detected.

Another type of non-ferrous metal detector is based on conductivity measurement. The detector produces electromagnetic waves that pass through the

Table 7.8 VS-50 antipersonnel mine. Reproduced from *Hidden Killers 1994: The Global Landmine Crisis*, P. F. Schultz III, Editor, Publication No. 10225, US Department of State, Bureau of Political-Military Affairs, 1994

General

Mine type	Antipersonnel
Lethal effect	Blast
Manufacturing country	Egypt, Italy, Singapore

Characteristics

Shape	Cylindrical
Height (mm)	45
Diameter (mm)	90
Total weight (kg)	0.185

Mine case

Material	Plastic
Number of fuse wells	0, integral fuse
Color	Variety of colors available including olive drab and sand

Mine fuse

Model	Integral
Type	Pressure
Actuation force (kg)	10

Explosive components

Main explosive type	RDX
Main explosive weight (kg)	0.043

Mine performance

Effective range (m)	Limited (typical with contact-fused, blast-effect mines)
Detectability	Difficult with hand-held metallic detectors (pressure plate is reinforced with non-metallic metal)
Antidisturbance	Possible (especially if the 'look-a-like' version, VS-50 AR is employed)

subsurface, causing induction of eddy currents within the mine. The intensity and phase of those eddy currents is a function of ground conductivity. Buried items and/or disturbed soil have conductivities different from the surrounding natural soil. The conductivity meter will detect those differences in conductivity. Mines have been characterized by measuring the frequency dependence of magnetic fields caused by electric currents induced in the target [256]. The frequency responses were measured by using a fixture incorporating a solenoid excitation coil, a receiving coil wound as a gradiometer and a

Table 7.9 PP-MI-SR antipersonnel mine. Reproduced from *Hidden Killers 1994: The Global Landmine Crisis*, P. F. Schultz III, Editor, Publication No. 10225, US Department of State, Bureau of Political-Military Affairs, 1994

General

Mine type	Antipersonnel
Lethal effect	Bounding fragmentation
Manufacturing country	Czechoslovakia

Characteristics

Shape	Cylindrical
Height (mm)	152
Diameter (mm)	102
Total weight (kg)	3.2

Mine case

Material	Steel, plastic
Number of fuse wells	1
Color	Olive drab

Mine fuse

Model	RO-1, RO-8
Type	Trip wire, pressure
Actuation force (kg)	4–8, 3–6

Explosive components

Main explosive type	TNT
Main explosive weight (kg)	0.362

Mine performance

Effective range (m)	20
Detectability	Readily detectable due to metallic case and fragmentations
Antidisturbance	Possible (however, no secondary fuse well, or AD features)

spectrum analyzer. The drawback of metal detectors is that they will not react to plastic, wood or cardboard mines. In addition, their use is complicated by the presence of other types of metal debris.

7.3.1.4 Magnetometers [257]

Magnetometers are used only for mines containing ferrous metal. These sensors are passive, they do not radiate any energy, but only measure the disturbance of the earth's natural geomagnetic field (which has a field strength of about 0.5 Gauss), by the metal-containing mine. Magnetometers must be sensitive enough to measure the weaker secondary magnetic field caused by a

Table 7.10 MON-200 antipersonnel mine. Reproduced from *Hidden Killers 1994: The Global Landmine Crisis*, P. F. Schultz III, Editor, Publication No. 10225, US Department of State, Bureau of Political-Military Affairs, 1994

General	
Mine type	Antipersonnel
Lethal effect	Directed fragmentation
Manufacturing country	Former Soviet Union
Characteristics	
Shape	Cylindrical
Height (mm)	434
Diameter (mm)	130
Total weight (kg)	25
Mine case	
Material	Metal
Number of fuse wells	1
Color	Olive drab
Mine fuse	
Model	MUV-type
Type	Trip wire, command detonation
Actuation force (kg)	2–5
Explosive components	
Main explosive type	TNT
Main explosive weight (kg)	12
Mine performance	
Effective range (m)	200
Detectability	Visually detectable
Antidisturbance	Possible (but not likely)

buried mine superimposed on the much larger geomagnetic background. Some magnetometers use two magnetic sensors, configured to measure the difference in magnetic field over a fixed distance; they are called gradiometers, as they measure the gradient of the magnetic field. Three types of magnetometers are currently being used:

- Fluxgate magnetometers measure the magnitude and direction of the magnetic field. By using an exciting coil to drive a highly permeable metal core in and out of a condition of saturation, magnetic flux lines in the core area are pulled into or out of the core. At saturation, the core inductance falls rapidly and current spikes are induced, which are detected by a separate sensing coil. Appropriate circuitry will measure the relative phase, polarity and size of the magnetic field. These magnetometers are

Table 7.11 PMN antipersonnel mine. Reproduced from *Hidden Killers 1994: The Global Landmine Crisis*, P. F. Schultz III, Editor, Publication No. 10225, US Department of State, Bureau of Political-Military Affairs, 1994

General	
Mine type	Antipersonnel
Lethal effect	Blast
Manufacturing country	Former Soviet Union, Iraq
Characteristics	
Shape	Cylindrical
Height (mm)	56
Diameter (mm)	112
Total weight (kg)	0.55
Mine case	
Material	Bakelite body with rubber cover
Number of fuse wells	0, integral fuse
Color	Black body with sand or black cover
Mine fuse	
Model	Integral
Type	Delay-armed, pressure
Actuation force (kg)	5–8
Explosive components	
Main explosive type	Trotyl
Main explosive weight (kg)	0.200
Mine performance	
Effective range (m)	Limited (typical with contact-fused, blast-effect mines)
Detectability	Readily detectable due to-fair amount of metallic content in fuse assembly and cover retainer
Antidisturbance	Possible (however, no secondary fuse well, or AD features)

inexpensive, reliable, rugged, and have low energy consumption. They can detect small items to a depth of 2–3 m. However, they are also sensitive to small fragments, and do not always discriminate between small, shallow fragments and deeper, larger intact munitions.

• Proton precession magnetometers are based on the principle that magnetic fields can be evaluated by measuring the movement of protons in a liquid, such as water, kerosene or another hydrocarbon. The liquid is placed in a container within a solenoid, the axis of which is aligned at right angles to the magnetic field under investigation. A polarizing field is developed when the solenoid is energized. This aligns the magnetic spin axis of the

Table 7.12 POMZ-2 antipersonnel mine. Reproduced from *Hidden Killers 1994: The Global Landmine Crisis*, P. F. Schultz III, Editor, Publication No. 10225, US Department of State, Bureau of Political-Military Affairs, 1994

General	
Mine type	Antipersonnel
Lethal effect	Fragmentation
Manufacturing country	Former Soviet Union, North Korea
	Former East Germany, China (POMZ-2 copies)
Characteristics	
Shape	Cylindrical
Height (mm)	107
Diameter (mm)	60
Total weight (kg)	1.77 (POMZ-2M)
	2.3 (POMZ-2)
Mine case	
Material	Metal
Number of fuse wells	1
Color	Olive drab
Mine fuse	
Model	MUV-type fuse
Type	Trip wire
Actuation force (kg)	2–5
Explosive components	
Main explosive type	TNT
Main explosive weight (kg)	0.075
Mine performance	
Effective range (m)	4
Detectability	Visually detectable
Antidisturbance	Possible (however, no secondary fuse well, or AD features)

protons in the liquid perpendicular to that of the magnetic field. When the polarizing field is instantaneously removed, the protons begin to precess about the axis of the ambient magnetic field. The frequency of precession will deviate from their natural frequency in proportion to the strength of the ambient field. A voltage, proportional to the magnetic field strength, will be induced in the solenoid. This type of magnetometer is more sensitive than a fluxgate magnetometer, but slower to use. It can detect munition items to a depth of 2–3 m.

- Optically pumped atomic magnetometers operate similarly to proton precession magnetometers, except that the proton is replaced by an atom of a

Table 7.13 PMD-6 antipersonnel mine. Reproduced from *Hidden Killers 1994: The Global Landmine Crisis*, P. F. Schultz III, Editor, Publication No. 10225, US Department of State, Bureau of Political-Military Affairs, 1994

General	
Mine type	Antipersonnel
Lethal effect	Blast
Manufacturing country	Former Soviet Union
Characteristics	
Shape	Rectangular
Length (mm)	196
Width (mm)	87
Height (mm)	50
Total weight (kg)	0.4
Mine case	
Material	Wood
Number of fuse wells	1
Color	Natural wood
Mine fuse	
Model	MUV-type fuse
Type	Pressure
Actuation force (kg)	1–10
Explosive components	
Main explosive type	TNT
Main explosive weight (kg)	0.200
Mine performance	
Effective range (m)	Limited (typical with contact-fused, blast-effect mines)
Detectability	Detectable due to metallic content of MUV-type fuse and detonator
Antidisturbance	Possible (however, no secondary fuse well, or AD features)

specific gas vapor, such as cesium or potassium. The light from a cesium or potassium vapor lamp is circularly polarized and directed along the approximate axis of the magnetic field to be measured. Light passes through an absorption cell containing vapor of the same metal. The intensity of the emerging beam, monitored by a photocell, is indirectly proportional to the strength of the magnetic field. Atomic magnetometers are more sensitive and have faster sampling rates than proton precession magnetometers.

Both metal detectors and magnetometers cannot differentiate between a mine and metallic debris. In most battlefields, the soil is contaminated by large quantities of shrapnel, metal scraps, cartridge cases, etc., leading to 100–1000 false alarms for each real mine. Each alarm means a waste of time and induces a loss of concentration.

Modern mines can have almost no metal parts, with the exception of the striker pin, for example. Although metal detectors can be tuned to be sensitive enough to detect these small items (current detectors can track 0.1 g metal at a depth of 10 cm), this may not always be practically feasible, as it will lead to the detection of smaller debris and increase the false alarm rate.

7.3.1.5 Mechanical detection techniques

Mechanical detection and detonation of mines involves heavy equipment, such as modified tanks, bulldozers, or trucks. It does not need sensors and is efficient on a suitable ground. Chains, attached on a rotating roller, hit the ground in order to blow up or destroy mines. On flat, hard ground, flails can be relatively effective against antipersonnel mines. But in some types of soil, the mines can be pushed in more deeply or flung around by the chains. Another possibility is to mount ploughs in front of a tank, which dig out the mines and move them away, mostly without exploding.

One man, operating a flail, can clear about $15\,000\,\text{m}^2$ per day.

7.3.1.6 Ground penetrating radar (GPR)

GPR works by emitting into the ground, through a wideband antenna, an electromagnetic wave, covering a large frequency band [258, 259]. This can be done using a short pulse or a pure sine wave whose frequency is varied continuously or by steps to cover the desired range. Reflections from the soil caused by dielectric variations, such as the presence of an object, are measured. By moving the antenna, it is possible to reconstruct an image representing a vertical slice of the soil. Further data processing allows horizontal slices or 3D representations to be displayed.

The high resolution needed to cope with the small objects encountered, enforces the use of a wide frequency band, thus limiting the penetration depth and increasing the image clutter. A center frequency of 1 GHz allows work at depths of 0.5–1 m in most soils with a resolution of a few centimeters. Penetration depth will also depend on the nature of the soils, having different attenuations. For example, desert sand has an attenuation of 1 dB/m for a frequency of 1 GHz, while clay has an attenuation of 100 dB/m at the same frequency. The GPR is able to detect nonmetallic materials as long as their characteristics are sufficiently different from the surrounding media.

A smaller type of sensor, the micropower impulse radar (MIR), uses a small antenna footprint ($<50\,\text{cm}^2$) and should allow a faster and simplified scan of the minefield. It emits about 2×10^6 pulses per second, each pulse has a width of the order of a nanosecond. With pulses so short, the MIR operates across a wider band of frequencies than a conventional radar, providing high resolution and accuracy, and making it less susceptible to interference from other radars. The power requirements are low, the emitted energy level being in the microwatt range.

Another approach with GPR is to look for complex resonances, specific to each target type, in the spectrum of the reflected signal.

A low weight radar has been mounted on a remotely controlled vehicle, capable of operating unharmed in adverse terrain.

GPR systems have been integrated with metal detectors, allowing mine detection in virtually all soil and weather conditions.

A vehicle-mounted stand-off mine detection system, which can detect and identify mine types at a range of up to 30 m, has been developed [260]. The system uses a stepped continuous wave signal, with three horn antennas. The range of signal frequencies, 0.5–4 GHz, allows the system to excite the resonant frequencies of all mine targets and identify them. The three horn antennas allow for azimuth detection, as well as range determination and identification. The total radiated power is 1 W.

7.3.1.7 Infrared (IR) imaging

Mines retain or release heat at a different rate from their surroundings, and during natural temperature variations of the environment, it is possible, using IR cameras, to measure the thermal contrast between the soil over a buried mine and the soil around it [258].

When this contrast is due only to the presence of the buried mine (alteration of the heat flow), a 'volume effect' is obtained. When it is due primarily to the disturbed soil layer above and around the mine (resulting from the burying operation), a 'surface effect' is obtained, which can be detectable for weeks after burial, and enhances the mine's signature.

Sensitive cameras with sufficient spatial resolution have to be used. The method is effective for maximum mine burial depth of 10–15 cm. In addition, results obtained with passive IR imagers can depend quite heavily on the environmental conditions. Also, there are cross-over periods (in the evening and in the morning), when the thermal contrast is negligible and the mines undetectable. Foliage is an additional problem.

A concept of a modular multisensor system for use on an airborne platform has been suggested [261, 262]. The sensor system comprises two high resolution IR sensors, working in the mid and far IR spectral regions, an RGB (red–green–blue color model) video camera with its sensitivity extended to the near IR in connection with a laser illuminator, and a radar with a spatial resolution adapted to the expected mine sizes. The system has an on-board real-time image processing capability and is planned to operate autonomously with a data link to a mobile groundstation. Data from a navigation unit is used to transform the location of identified mines into a geodetic coordinate system.

7.3.1.8 Passive millimeter wave (MMW) detection [258]

In the millimeter wave band, soil has a high emissivity and low reflectivity. Conversely, metal has a low emissivity and strong emissivity. Soil radiation depends therefore almost entirely on its temperature, while metal reflection depends mostly on the low level radiation from the sky. It is possible to measure this contrast using a millimeter wave radiometer device. Tests in ideal laboratory conditions have demonstrated the capability of detecting metallic objects buried under 3 in of dry sand, working at 44 GHz. At this frequency, even water content of a few percent results in a very poor penetration depth.

Tests have been carried out subsequently also on plastic targets, which produce a much smaller delta T than the metal ones (they have much lower reflectivity and transparency to radiation rising from below them), working at 44 and 12 GHz, and recently also at 5 GHz. The trend towards lower frequencies present the advantage of increased penetration, especially in most soil, at the obvious price of some loss in spatial resolution.

Passive MMW radiometers are simpler devices than GPR. They should suffer less from clutter problems and can be used to generate 2D images of objects placed on the surface (possibly under light vegetation) or shallowly buried (a few centimeters), with best results in dry soils, and for metallic targets.

7.3.1.9 Acoustics [258]

Ultrasound detection consists of the emission of a sound wave with a frequency higher than 20 kHz into a medium. This sound wave will be reflected on boundaries between materials with different acoustical properties. Such systems should be capable of good penetration through very wet and heavy ground such as clay, which makes them somewhat complimentary to GPR.

Experimental research has been conducted in the laboratory on the use of ultrasound impulse echo techniques for antipersonnel mine detection in the framework of a simulation of mines thrown into rice fields (i.e. under water). A 15 MHz probe and a scanning step of 0.6 mm (along x, y) were used. Some signal processing methods and pattern recognition methods have been implemented to discriminate between mine-like objects and other objects. At such high frequencies ultrasound does practically not penetrate soil, which is the reason why such tests were targeted at finding mines in water.

7.3.1.10 Backscattered X-rays

The interaction of photons with matter at energies below 1 MeV is dominated by the photoelectric effect and Compton scattering (see paragraph 3.3.1.4).

The technique relies on the absorption and scattering of energetic photons. A detector system located above the soil intercepts backscattered photons. At X-ray energies below 100 keV, Compton scattering depends weakly on the Z of the scattering material, while the photoelectric absorption varies as Z^4. The average Z of the soil is $Z_{soil} = 12$–14, while the average Z of explosives and plastic is $Z_{exp} = 6$–7. It is therefore possible to detect shallow-buried plastic mines by detecting back-scattered X-rays: while the soil will absorb most X-rays impinging upon it, the plastic mine will back-scatter a large fraction. Thus the mine appears as a luminous spot on the dark background of the soil [254]. Metallic mines will absorb even more efficiently than soil, and can therefore not be differentiated from it.

An imaging system can be obtained by using an array of detectors in a square configuration. The system can image land mines down to about 4 in through water and snow and all types of debris, such as rocks, logs and leaves. It can image 1 m^2 in about 5 min.

Such a system successfully imaged antitank mines buried in sand and rocky soil. The image showed enough detail to ascertain the type of mine and the location of its fuse, which is important for unearthing it.

Lateral migration radiography (LMR), a form of Compton backscatter radiography, has been applied to the detection and identification of landmines [263]. The LMR system consists of two inner uncollimated detectors, positioned to optimally detect first scattered photons, and two outer collimated detectors, designed to detect primarily photons that have had two or more scatterings. The difference between the collimated and uncollimated detector response to both the landmines themselves and the different types of landmine image masking phenomena, form the basis of the image enhancement and landmine identification procedures. The primary component of the uncollimated detector response is surface feature information, while the collimated detector signal contains information about both the surface features and the buried objects. Detector height variations, due to surface irregularities, can be eliminated by using the information from both detector systems.

7.3.1.11 *Multi-sensor mine detectors (sensor fusion)*

Several mine detectors use multiple sensors. For example, a vehicle-mounted mine detector has been designed to detect landmines using a multisensor system mounted on a remote controlled vehicle [264]. The system has the ability to detect antipersonnel and antitank mines, using a ground penetrating radar (GPR) for close-in detection and IR and UV sensors for standoff detection.

The GPR close-in sensor, which consists of a technology combination of 3D processing and advanced frequency stepped radar, detects and identifies buried landmines ≥ 2 in in diameter off-road and at least 8 in in diameter on-road. In addition to the sensors, the system also has video cameras, a Global

Positioning System to determine locations, remote controlled paint sprayers for marking and an operator's command station.

A multisensor, vehicle-mounted, teleoperated mine detector consisting of a forward-looking infrared imager, a 3 m wide down-looking highly sensitive electromagnetic induction (EMI) detector and a 3 m wide down-looking ground penetrating radar (GPR), which all scan the ground in front of the vehicle, has been suggested [265]. Scanning sensor information is combined using a suite of navigation sensors and custom designed navigation, spatial correspondence, and data fusion algorithms. Suspect targets are then confirmed by a thermal neutron analysis (TNA) detector. A major condition for data fusion is that the sensors must have independent outputs, so that multiple alarms due to a target would be indicative of a mine, even though any single detector could occasionally produce a false alarm.

A hybrid remote-sensing method using high-power microwave (2.45 GHz, 5 kW) illumination and passive infrared detection has been developed for the detection of shallow buried landmines [266]. The thermal signature of the mine at the soil surface was detected in the 8–12 μm region, both in near real-time as well as after a brief time-delay following illumination. The thermal signature at the soil surface is primarily made up of two components: one component, due to the interference of the incident beam and the beam reflected by the mine, and a second component, generated by the variations in heating due to differential microwave absorption by the mine and the surrounding soil, resulting in a temperature variation at the soil surface above the mine relative to the nearby soil surface. Both signatures are dependent on the complex dielectric constants of the mine and the soil.

7.3.2 Chemical sensing of explosives

In contrast to anomaly detectors, which look for the container holding the explosive, chemical sensors look for the explosives contained inside the mines, either by sniffing the explosive vapor or by bulk detection.

7.3.2.1 Vapor sniffing detectors

When a landmine is buried in the ground, vapors emanating from the explosive begin to leak from the landmine into the soil. Eventually, a fraction of these explosive vapors reach the surface of the ground and escape into the air over the landmine.

The fate of the explosives molecules, once released from the landmine, and the mechanism by which they are transported through the soil to the surface of the ground is a complex process. The movement of explosives molecules through soil, air, and water is influenced by the type of soil, as well as by environmental conditions, such as temperature and rainfall. Water and soil pose different problems. Explosives dissolve easily in water, making it difficult to extract the explosives molecules. Water and soil are often a dirty environment, with salts, organic materials and pollution, interfering with the detection process [267]. As explosives commonly used in landmines have very low vapor pressures, the concentration of explosive vapor emanating from the landmine is very low. In addition, explosives interact strongly with a variety of surfaces. These interactions could lead to adsorption of explosives on soil particles. Other processes, such as microbial degradation [11] might change the chemical composition of the explosive compounds as they diffuse through the soil, giving rise to degradation products. The result is that the concentration of the explosive vapor in the air over a landmine is very low.

A one-dimensional model, developed for screening agricultural pesticides, was modified and used to simulate the appearance of a surface flux above a buried landmine, estimate the subsurface total concentration, and show the concentration at the ground surface [268–270]. There is a strong correlation between the movement of explosive signature molecules in the near subsurface soil and that of applied agricultural pesticides, as both can be organic molecules having a low vapor pressure. Figure 7.3 shows a conceptual model of the environmental fate and transport processes that impact

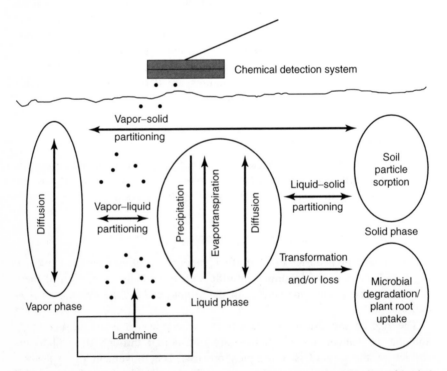

Figure 7.3 Environmental fate and transport model for chemical detection of buried landmines. Reproduced from Phelan, J. M., et al., Sandia National Laboratories, Report No. SAND97-1426, 1997, with permission.

the movement of landmine chemical constituents to the surface for chemical detection. Chemical vapors emanate from a buried landmine by permeation through plastic case materials or through seals and seams, and from the initial surface contamination of the case. Vapor phase diffusion transports molecules away from the landmine. The vapors may partition into the aqueous phase of the soil water which may then be transported to the surface through advection, driven by evapotranspiration or to the depth by precipitation, and through diffusion driven by concentration gradients. Under extremely dry soil conditions near the ground surface, vapor phases may be directly sorbed to soil particles. When in the liquid phase, chemicals may also sorb to soil particles. Soil particle sorption can be considered a temporary storage reservoir for the explosive constituents, where they may be released under reversible partitioning reactions, but some of the explosives may also permanently bond through chemisorption reactions. Transformation and loss of explosive constituents also occurs during microbial degradation and uptake by the roots of certain plant species.

7.3.2.1.1 Trace analysis of explosive vapors emanating from mines

The concept of using explosives vapor detection to locate mines has been under investigation for many years [271–275]. Characterization of a mine should include not only the vapor of the explosive itself, but also vapors of impurities present in the explosive. For example, in 2,4,6-TNT, small amounts of other isomeric forms of TNT as well as several isomers of dinitrotoluene (DNT) are usually present [272]. TNT isomer impurities included 2,3,4- and 2,3,5-TNT, and DNT isomers included 2,4-, 2,5-, and 3,5-DNT. The concentration of each of the DNT and TNT isomeric impurities in the solid TNT was less than 0.1%, and for several impurities, less than 0.01%. As the vapor pressure of some of the DNT impurities are much higher than those of TNT itself, it could be useful to characterize a mine by detecting the vapor of one of those impurities.

Cyclohexanone was identified as a major constituent of the vapor evolved from the explosive Composition B (found in antitank mines of type M15 and M19) [274, 275]. Cyclohexanone is not a component of the explosive itself, but is a solvent used in the recrystallization of RDX, a major component of Composition B. Large amounts of acetone were found in the vapors evolving from M19 mines and M19 casings. Acetone probably originates from the materials used in the construction of M19 nonmetallic casings.

7.3.2.1.2 Use of dogs

Dogs have been used to sniff explosive vapors emanating from buried mines. Under handler control, the dogs operate in suspected mine areas and alert the handler that they have detected a mine or tripwire by sitting within 1 m of the detection. The dog/handler relationship is very important: handlers must have the ability to sense their dog's abnormal behavior that could possibly indicate that a mine may be present. Dogs are very sensititve but have some limitations: they tire easily, work inconsistently and require much time and effort to train properly. Their 'useful life' is also relatively short. Localization accuracy is usually not very good (several meters), because explosives vapors can penetrate the ground and the vegetation in an area up to 10 m from the real location of the mine after some months, and that trace particles might also be scattered around. The mine's vapor release rate can also change significantly over time after burial.

Precise location is not a problem when verifying with dogs vast stretches of land, in order to save precious time by concentrating on areas which really need to be demined.

Although it has been shown that dogs can detect vapor compounds at very low concentrations, the chemical basis for canine explosives detection remains largely unknown. A study has been carried out to determine what are the compounds dogs learn to use in recognizing explosives used in landmines [276]. This was accomplished by training dogs under behavioral laboratory conditions to respond differentially on separate levers to blank air, a target odor (i.e. an explosive), and all other nontarget odors. Vapor samples were generated by a serial dilution vapor generator, whose operation and output was characterized by GC/MS. The study was conducted using TNT and C-4.

Dogs are trained using substances similar to those they will be expected to detect in the field. Most of the substances that dogs are trained to detect are mixtures of a variety of compounds. The vapor from these substances may not contain the same proportions of each compound as in the original substance. Canine olfactory systems are capable of reacting to a number of compounds in explosives and other target substances. With sufficient training, the compound or compounds whose detection most often results in reinforcement become what has been termed the detection odor signature. This is not necessary the chemical signature of the explosive vapor, but rather a behavioral detection signature. It was found that the detection odor signature for NG smokeless powder does not include NG, but acetone, limonene, and toluene, while the detection odor signature for C-4 did not include RDX, but cyclohexanone and 2-ethyl-1-hexanol [276].

7.3.2.1.3 Artificial sensors

Advances in the understanding of olfaction are leading to artificial electronic noses based on an array of sensors that bind airborne molecules with only modest specificity. Detection is based on unique patterns of responses, generated by the sensor array in the presence of an odor.

A low-power, broadly responsive thin-film resistance vapor sensor has been developed [277, 278]. Carbon black–organic polymer composites have been shown to swell reversibly upon exposure to vapors. Thin films of carbon black–organic polymer composites have been deposited across two metallic leads, with swelling-induced resistance changes of the films signaling the presence of vapors. To identify and classify vapors, arrays of such vapor-sensing elements have been constructed. These arrays contain the same carbon black conducting phase, but different organic polymers as the insulating phase. The variations in gas–solid partition coefficients for the various polymers of the sensor array, produce a pattern of resistance changes that can be used to classify vapors and vapor mixtures, including explosives.

An optical fiber sensor array containing a variety of differentially reactive sensors has been developed [279–282]. When an analyte is presented in a pulsative form, each sensor, which is based on a polymer/dye combination, produces a unique fluorescence vs. time signature. The system uses neural network analysis with pattern recognition, to discriminate between the various compounds.

Individual sensors are made using a solvatochromic fluorescent dye in combination with one or more of the following: polymer(s), micrometer-sized beads, or various microsized particles, such as alumina particles. The polymers and additives can act as concentrators and/or adsorbents of organic vapors which help to maximize analyte/dye interaction. The sensors can be chemically fixated onto coverslips or onto the distal end of optical fibers for testing analyte vapors, delivered in a pulsative mode. Individual optical sensors are analyzed according to their change in fluorescence and/or shift in wavelength, resulting from a polarity change due to the analyte vapor presentation. Parameters which contribute to the optical response giving each sensor a unique chemical signature for a particular analyte vapor are: vapor diffusion through the polymer layer, polymer type, polymer layer swelling, analyte vapor adsorption, pulse time, and flow rate.

The designed 'artificial nose' vapor detection system [279–282] uses an array of cross-reactive nonspecific sensors to make it resemble the olfactory system. Multiple sensors on a single fiber were found to respond differently to different analytes. Consequently, thousands of individual bead sensors were incorporated into a coherent imaging fiber of 1 mm diameter. Etching the cores of about 20 600 individual pixels at the distal end of a fiber, created tiny wells, into which polymeric beads with different encoding chemistries were inserted in order to create an array consisting of thousands of sensors. Figure 7.4 shows a schematic representation of the odor-sensing process [279]. Vacuum-controlled pulses of various air/solvent vapor dilutions are delivered directly to the distal ends of fibers in the sensor bundle. Exposure to vapor induces changes in fluorescence intensity at a given wavelength which are then recorded and plotted versus time to produce a temporal response pattern for each sensing site to a given odor. Selective and sensitive optical sensors were developed for TNT and DNT. Preliminary experiments, using cover slips with 3 μm diameter porous silica beads, dyed with Nile red, showed that beads responded differently to vapors of 1,3-dinitrobenzene (DNB), 4-nitrotoluene, and DNT. The imaging system consisted of an inverted microscope couple to a CCD detector, which allowed discrimination between each sensor response.

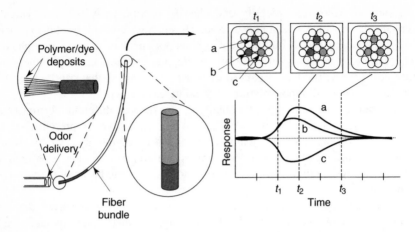

Figure 7.4 Schematic presentation of the odor-sensing process. Reprinted with permission of author and publisher from Dickinson, T. A., et al., *Nature*, **382**, 697, © 1996 Macmillan Magazine Limited.

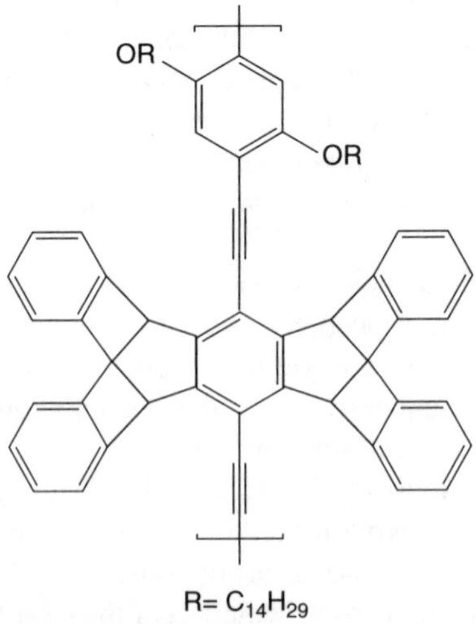

$$R= C_{14}H_{29}$$

Figure 7.5 Structure of the pentiptycene polymer. Reprinted with permission from Yang, J.-S., et al., *J. Am. Chem. Soc.*, **120**, 5321, © 1998 American Chemical Society.

Thin films of the pentiptycene polymer, the structure of which is shown in Figure 7.5, were found to display high fluorescence quantum yield and stability for the vapors of TNT and DNT [283]. Figure 7.6 shows the time-dependent fluorescence intensity of the polymer upon exposure to TNT. The highest intensity is obtained at 0 s, and lowest one at 600 s. The fluorescence quenching increases to 50% within 30 s and to 75% at 60 s (Figure 7.6, inset).

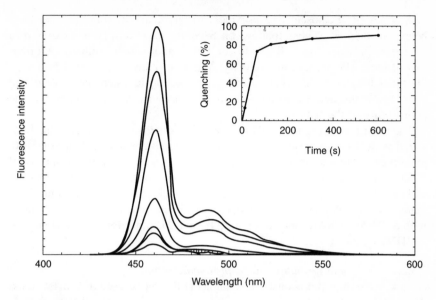

Figure 7.6 Time-dependent fluorescence intensity of the polymer upon exposure to TNT vapor at 0, 10, 30, 60, 120, 180, 300, and 600 s (top to bottom) and the fluorescence quenching as function of time (inset). Reprinted with permission from Yang, J.-S., et al., *J. Am. Chem. Soc.*, **120**, 5321, © 1998 American Chemical Society.

The mechanism of fluorescence attenuation can be attributed to the electron transfer from the excited polymer to the electron acceptor TNT.

7.3.2.1.4 Polymer coatings for chemical sensors

By choosing an appropriate sorbent coating for chemical sensors, it is possible to design sensors for specific groups of analytes [284]. The properties of coatings have been specifically evaluated for application on surface acoustic wave (SAW) devices (see paragraph 2.4.3).

The acoustic wave generated is very sensitive to material at the crystal surface, and coating a thin sorbent film on the surface converts the SAW device into a highly sensitive chemical sensor. The sorbent coating is usually a polymeric material that absorbs and concentrates vapor molecules in the region where the acoustic wave propagates. For fast vapor absorption, polymers with high permeability are required. In addition the polymer must be nonvolatile, and if applied to the SAW device by spray coating, it must be soluble in a volatile solvent and exhibit stable, well-formed coatings on the surface of the SAW device.

The chemical processes that govern the sensitivity and selectivity of a polymer-coated SAW device are very similar to those of the solution process of a vapor in a liquid solvent. When a solvent is exposed to a vapor, vapor molecules distribute themselves between the gas and liquid phases, and a thermodynamic equilibrium is established [284].

A series of chemoselective polymers to target nitroaromatic vapors have been evaluated as sorbent coatings on SAW devices [285]. When exposed to a vapor, a polymer coated SAW device sorbs vapor which results in perturbation of the SAW propagation velocity. Vapor sorption is monitored as a shift in signal frequency. The nature of the interaction between the coating and vapor molecules determines the selectivity, sensitivity, signal kinetics, and the reversibility of the sensor.

The most promising materials tested included siloxane polymers, functionalized with acidic pendant groups, that are complimentary in their solubility properties to nitroaromatic compounds. A series of hexafluoroisopropanol functionalized aromatic siloxane polymers were synthesized, characterized, and evaluated as vapor sorptive coatings for use with SAW devices for nitroaromatic compounds. The new polymers exhibited detection limits, obtained with a 250 MHz SAW device, of 3 p.p.b. and 235 p.p.t. for nitrobenzene and 2,4-DNT, respectively.

7.3.2.1.5 Thin-film resonators for explosive detection

A bulk wave thin-film acoustic resonator (TFR) can be used as a mass transducer when combined with an appropriate sorbent coating and thus provide the basis for a TNT vapor detection scheme [286]. In this scheme, the sampled air will pass over an array of sensors, each of which has a different sorbent coating. The change in the signal from the different sensors will then be used to develop signature response patterns to identify the presence of the TNT vapor, and to differentiate between the TNT signature and the interferents.

Bulk-wave acoustic resonators (see also: surface acoustic wave, SAW, devices in paragraph 2.4.3) are electromechanical devices fabricated by depositing thin metal electrodes on the opposing faces of a piezoelectric slab. Resonance conditions are established when the thickness of the slab is equal to a multiple of half the acoustic wavelength. This resonance can be detected by the frequency of the voltage applied to the electrodes. When material is added to one of the electrode surfaces, the resonant path length increases and the acoustic frequency decreases. The change in resonance frequency is a linear function of the mass of the material adsorbed on the electrode surface.

In a bulk-wave TFR, the piezoelectric slab is fabricated as a thin film sandwiched between metal films. Since the mass sensitivity is inversely proportional to the resonator thickness, a thin-film resonator is expected to have a very high mass sensitivity. TFRs were made using AlN as the piezoelectric film. In order to achieve vapor selectivity and sensitivity, organic sorbent coatings of 20–100 nm thick were applied to the TFR.

Results showed that TFRs, operated at 2 GHz, were capable of detecting 5 p.p.b. TNT with a signal-to-noise ratio of 200 : 1.

7.3.2.1.6 Microbial mine detection

Various naturally occurring microbial species can be stimulated by nitrogen, TNT, DNT, nitrous oxide, and other chemical components found in explosive

materials [287]. Microorganisms, in the presence of an environmental contaminant, can adapt to the contaminant and utilize it as a nutrient source.
Likewise, microorganisms, in the presence of land mines, can adapt and
flourish on explosives. By using selective enrichment culturing techniques
and identifying unique microbial communities, it has been suggested [287] that
microbial microorganism strains which produce a bioluminescence reaction in
the presence of landmines could be used for detection.

Microbiological detection systems, established for other compounds, such
as trichloroethylene, have shown sensitivities in the sub-p.p.b. to upper p.p.t.
range.

7.3.2.1.7 Explosive sampling systems

A fate and transport study of TNT in soil predicts that over 90% of the explosive escaping from a landmine is adsorbed onto the soil solid phase particles,
about 10% remains in the liquid phase, and only a trace (about 10^{-6}) is in the
vapor phase [268]. Therefore, a small soil particle sample collected from the
surface of the ground, could easily contain a much larger quantity of explosive
than a large sample of air, immediately over the landmine. An electrostatic
particle sampler is being developed which is capable of removing a high
percentage of particulate matter from large quantities of air [288]. Soil particles on the surface of the ground are entrained in a stream of air and drawn into
the particle sampler. Once the particles are in the sampler, they become electrically charged under the influence of a strong electric field, and are collected
on an oppositely charged electrode. The electrostatic sampler consists of a
discharge electrode constructed from a small gauge wire suspended along the
axis of a cylindrical collection electrode. The dust laden air stream is drawn
through the space separating the electrodes. A potential difference of about
4.9–5.5 kV is applied across the two electrodes, generating an electric field
within the particle sampler, thus producing a corona discharge. The electrons formed are accelerated away from the discharge electrode and can be
captured by electronegative gas molecules such as oxygen, forming negative
molecular ions. These negative ions attach to neutral dust particles, forming
negatively charged dust particles, which deposit on the collection electrode.
The particles are then removed from the sampler and transferred to a desorp-
tion/preconcentration stage. In this part of the device the explosive compounds
are thermally desorbed from the soil particles and preconcentrated by cold
trapping. The concentrated explosive molecules are then analyzed by a vapor
detector, such as IMS.

A three-stage water sampling system has been designed [289]. The first stage
involves sampling the water near a suspected target. The second stage involves
separating and concentrating the explosive molecules from the water, and in
the third stage the explosive analyte is transferred to a detector for analysis.
A schematic of the system is shown in Figure 7.7. The sampling tube, which
contains an inlet hose and particulate filter, utilizers a pump to collect a sample
of water. The sample is passed through the concentrator which is filled with

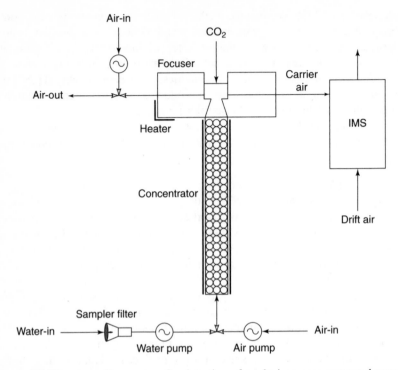

Figure 7.7 Water sampling system for detection of explosives near suspected targets. Reproduced from Chambers, W. B., et al., Conf. on Detection and Remediation Technologies for Mines and Minelike Targets III, Orlando, FL, 1998, with permission.

glass beads coated with a thin polymeric film to enhance adsorption of the explosive analyte. After the sample has passed through the concentrator, the concentrator is drained, and air is pumped through the column to remove excess water. The concentrator column is then heated to desorb the analyte, which is transported by a flow of air to the focusing stage of the concentrator. The focuser utilizes a cold surface to condense and trap the analyte. The surface of the cold trap is subsequently heated to desorb the analyte into the carrier air flow for detection by IMS.

A soil sampler has been developed, based on field and laboratory experiments with explosive contaminated soil [289]. It uses a light-weight probe for extracting and concentrating explosive vapor from soil in the vicinity of landmines. The nature and quantity of the chemical signature associated with buried landmines had first to be determined. It was asumed that the explosive signature flux from buried mines would be derived from two rates: short-term surface contamination, and long-term diffusion through plastic cases, seals and gaskets. The quantity of initial explosive contamination on the mine surfaces was determined by swiping the surface with solvent moistened swabs. Results of 2,4-DNT showed that for an AT-1 mine, with a bakelite casing,

$5.3 \, \text{ng/cm}^2$ of TNT and $29 \, \text{ng/cm}^2$ for 2,4-DNT were obtained, while RDX could not be determined. For an AT-4 mine, with a polyethylene casing, results were 4.3, 6.4 and $5.1 \, \text{ng/cm}^2$ for TNT, 2,4-DNT and RDX, respectively. Soil samples, taken 150 days after burying an AT-4 mine, contained $0.02–2.03 \, \mu\text{g/g}$ TNT and $0.01–2.70 \, \mu\text{g/g}$ 2.4-DNT.

As previously mentioned, the majority of the chemical signature will be found adsorbed to soil particles. The soil sampler utilizes a solid phase microextraction (SPME) method. In this method, water is added to the soil sample in a sealed septum vial and heated to 60°C, in order to increase the solubility and vapor pressure of the analytes. A fused silica fiber tip, coated with a chemical specific polymeric adsorbent, is inserted through the septum into the vapor 'headspace' to acquire the analytes, but can also be extended into the aqueous phase. The vapor or aqueous phase analyte is collected on the fiber coating. The fiber is then removed from the vial and thermally desorbed in the heated inlet of a GC or IMS. A field deployable soil/sediment probe is being designed, that can be coupled with a miniaturized IMS for detection of landmines [289]. In this design, hot water or steam is injected into the soil or sediment to create a heated, aqueous, 'microenvironment' in which the explosive molecules are extracted from the soil particles and preferentially re-adsorbed to the chemically selective solid phase coating of the probe. The probe is then retracted and the solid phase is thermally desorbed into a miniaturized IMS.

Another system for the sampling of traces of explosives from landmines is based on collection of samples by a vacuum, extraction of the residual explosives bound to the soil with a solvent under pressure and detection by IMS [290]. The system collected a sample by vacuuming and used a sifter device which selected small particles of soil. It then robotically extracted the residual explosives with solvent to weaken the explosive bounded to the soil. The soil sample was then desorbed and analyzed by IMS. Detection limits were 0.4 ± 0.1 p.p.b. w/w for TNT and 7.4 ± 1.9 p.p.b. w/w for RDX. Results showed that the concentration of explosives in the soil, shortly after the burial of mines, was below the detection limit of the system. It was found that the explosive level contained in the soil increased with time to 8 p.p.b. w/w for TNT 10 months after the burial of the mines.

7.3.2.1.8 Microelectromechanical systems (MEMS)
As has been previously mentioned, in the vicinity of landmines, some explosive particles can be found, depending on the conditions of the soil and environment. The median surface contamination of TNT and RDX of depot-stored mines has been found to be 15 and $1 \, \text{ng/cm}^2$, respectively. The most effective way to obtain soil particles for sampling is to accumulate them on a clean surface, concentrate them, and then analyze them and/or volatilize them for the vapor detector.

The basic idea of the MEMS method is to ultrasonically stimulate a target area, detaching explosive particles, and collect them [291]. They are then irradiated with selective infrared radiation and deflagrate with release of heat, which is detected by a bimetallic cantilever beam. The system uses focused megahertz frequency air-ultrasound waves to loosen and remove small particles ($<100\,\mu m$). The optimal frequency for particle stimulation is 1–5 MHz. In this range of frequencies, power densities of the order of $1\,kW/m^2$ should be sufficient to remove particles of the order of $10\,\mu m$ radius. The particles in the vicinity of the landmine are collected, by means of a mild vacuum pressure, onto an array of bimetallic cantilever beams. A source of optical radiative energy is supplied to the explosive particles in such a way that the sensor does not absorb the energy, but most of the energy is selectively absorbed by the explosive particles. The wavelength of the energy source is chosen to lie in the peak absorption spectrum of wavelengths of the explosives, so that all the energy is absorbed by the explosive. The top layer of the sensing cantilevers is a highly gold-coated reflecting surface in order to reflect all the optical energy in that range of wavelengths. They do not deflect the heat of the light source which is used to ignite the explosive, but will deflect the heat released by the explosive particles. The explosive particles, on absorbing the energy, deflagrate and release heat. The long wavelength heat thus released is absorbed by the cantilever beam through thermal conduction and the beam undergoes deflection. An optical set-up, which includes a helium–neon laser, measures the deflection of the cantilevers.

7.3.2.2 Bulk detection

7.3.2.2.1 Thermal neutron activation (TNA)

A thermal neutron activation (TNA) sensor, mounted behind a teleoperated, vehicle-mounted, multisensor mine detection system, has been used for confirmation of the presence of mines [292].

Neutrons from a $100\,\mu g$ ^{252}Cf (nominally 2.3×10^8 n/s) source within the TNA sensor penetrate the ground and interact with buried objects. These interactions result in $10.83\,MeV$ γ-ray signals associated with neutron capture of ^{14}N. The TNA sensor, by means of four $7.62 \times 7.62\,cm$ NaI detectors, will confirm high concentrations of nitrogen typically found in explosives. Among the gamma-rays emitted by ^{15}N, upon thermal neutron capture, the $10.83\,MeV$ gamma-rays were chosen because at this energy there will be virtually no competing reactions, except the weak $10.611\,MeV$ transition from neutron capture in ^{29}Si, a common constituent in most soils. Another advantage of this gamma-ray energy is that it is sufficiently isolated that low energy resolution NaI detectors can be used instead of lower efficiency, more expensive, and less rugged, high-resolution cryogenically-cooled intrinsic Ge detectors.

Once a target of interest has been determined by the multisensor system, its ground position is tracked until the confirmatory sensor, the TNA, is positioned within a 30 cm radius circle above the location. By examination of the activated gamma-ray spectrum for a preset count time at that location, the TNA will either confirm or deny the existence of a mine by the presence of bulk nitrogen.

Results showed that the TNA system was capable of confirmatory detection of landmines having nitrogen masses of greater than about 100 g, in a few minutes, over a radial area of at least 1200 cm^2. The system is thus capable of confirming the presence of the following:

- All surface-laid or shallowly-buried antitank mines in a few seconds to a minute (depending on the mass of the explosive).
- Antitank mines down to 20 cm depth in <5 min.
- Large (>100 g nitrogen) antipersonnel mines in <5 min.

The TNA sensor thus reduces the false alarm rate of the multi-sensor detection system to an acceptable overall rate.

7.3.2.2.2 Nuclear quadrupole resonance (NQR)

Nitrogen-14 nuclei possess an electric quadrupole moment since their charge distribution is not spherically symmetric. As a result, in crystalline structures, the electric fields generated by the particular molecular arrangements around the ^{14}N nuclei give rise to different sets of excited energy states.

In explosives such as RDX and TNT, transitions among such states cause characteristic RF absorption spectra that can uniquely identify the crystalline material. The splittings between these energy states is of the order of 10^{-8} to 10^{-5} eV, with corresponding RF frequencies in the range 10–1000 MHz. By rotating the ^{14}N nucleus with an external oscillating magnetic field that couples to the nitrogen nuclear magnetic moment at such frequencies, the nucleus can be made to oscillate between two such excited energy states, giving rise to the absorption line spectra, which is characteristic of the particular crystalline structure. A full description of NQR spectroscopy can be found in paragraph 3.3.3.

Although NQR has been successfully tested to detect explosives in airport luggage, in order to adapt it to mine detection, several difficulties must be overcome [254]: the detector cannot surround the mine, therefore a one-sided geometry must be used. This will probably reduce the signal, requiring longer integration times. A major difficulty to overcome is the detection of TNT, which is the explosive ingredient of 80% of the landmines. TNT produces a weak signal in NQR, so that it will probably require again longer integration times, as well as improvements in coil geometry and detector electronics.

A NQR system for the detection of landmines, which tries to overcome these problems, has been designed [131, 140, 293]. The design includes two basic developments: a detection coil suitable for probing the ground for landmines buried at typical depths, and an increase in the NQR signal obtained from TNT. A transmitting and receiving antenna system is required that is capable of projecting the RF excitation field to sufficient depth and then detecting the induced NQR signal. This antenna must operate over all types of ground materials with variable standoff distance and tolerate high levels of RF noise from the environment. The antenna used consisted of a lower spiral coil to perform the NQR measurement, while an identical upper coil was used to cancel the background magnetic fields. For the NQR measurement, the antenna operates almost entirely as a magnetometer due to the rapid decrease of NQR signal with increasing distance above the ground. For background RF interference, the antenna operates as a gradiometer.

As has already been mentioned, TNT is more difficult to detect than other explosives because the energy of its NQR response is divided among 12 spectral lines. In addition each line can be interrogated for approximately 100 ms before the signal becomes unmeasurable. As the NQR signal amplitude is approximately inversely proportional to the pulse interval, τ, over the range measured (0.76 ms $< \tau <$ 1.52 ms), the pulse interval was reduced by incorporating an advanced two-stage Q damping system [293]. This Q damper is able to reduce the ringdown time of the highly resonant NQR antenna from around 500 μs to 50 μs, and the pulse repetition delay to below 500 μs. As the majority of the NQR signal can be obtained from a few of the spectral lines, rapid frequency switching hardware was developed in order to detect multiple spectral lines in rapid succession. Results showed that the single-sided antenna was within a factor of four of having sufficient sensitivity to detect the quantities of RDX found in the smallest known antipersonnel mines. The results for TNT showed enough sensitivity to detect the TNT charge only in antitank mines.

7.3.3 Neutralization of landmines

7.3.3.1 Laser neutralization [294]

The principle of mine and ordnance neutralization by laser is based on heating of the mine or munition case until the explosive filler ignites and starts to burn. If it is a metal case, the heat is conducted through the case and target irradiation is continued until the temperature of the inside wall and the temperature of the explosive filler exceeds its combustion temperature. If it is a plastic case, the case is irradiated until it has been penetrated and the explosive filler is ignited, either directly from the laser radiation or from the flames burning the plastic case. Explosive filler continues to burn until a fuse or detonator is set off. The system, mounted on an armored vehicle, consists of a visible color camera for scene display and a joy stick for target scene scanning and pointing. It utilizes a diode pumped Nd–YAG laser for neutralizing the landmine. The invisible Nd–YAG laser is co-boresighted with a visible green doubled Nd–YAG tag laser. The operator uses the joystick to slew the visible camera, which has a large zoom lens, to perform surveillance of the scene. Once a target is visibly detected, the operator slews the green tag laser onto the target and selects an aimpoint. Then, the high power laser is turned on and the target is neutralized. The system can neutralize targets at ranges of up to 200–300 m, as long as the targets are in a line-of-sight orientation.

7.3.3.2 Foam neutralization

A mine marking and neutraliztion foam has been developed, which provides a method for safely marking and removing landmines, while rendering the fuse inoperable, if in-place neutralization is not possible [295]. The foam is a two-part, hand dispensed, polyurethane foam that is applied to exposed mines. The bright color provides a quick and easy method of marking mines and dangerous areas. As the foam hardens, it impregnates the exposed parts of the mine and renders the fuse inoperative. The hardened foam prevents detonation of the mine if a deminer accidently steps or falls on the marked mine. The foam also acts as an adhesive to glue any wires attached to the exposed mines, suspected to be booby-trapped. The foam operates over a temperature range of 0–40°C, and is environmentally inert.

Another foam, which is a nitromethane based explosive, has been used as blasting agent to destroy mines in place. It will neutralize individual mines and some unexploded ordnance [295].

7.3.4 Performance evaluation of mine detection techniques

Many technologies in use for the detection of landmines and unexploded ordnance (UXO) suffer from high false-alarm rates, even at modest probabilities of detection. The largest factor contributing to poor detection rates and high false-alarm rates for anomaly mine detection systems is clutter [296]. The source of this clutter can be either naturally occurring or man-made.

In order to determine the source of part of these sensor response anomalies, an excavation was carried out [297]. The goal was to catalog sources of clutter signatures observed during the earlier sensor data collection effort, including sensor types such as magnetometer, infrared (IR), electromagnetic induction (EMI), and ground penetrating radar (GPR). Preliminary analyses of the data showed a substantial number of sensor responses that appeared to be similar to the response expected from buried landmines.

The objects found at the site included metal banding (64%), metal scrap, car parts, communications wire, training rounds, metal plates, two fence posts, a soda can, a piece of foil, a 55 gallon drum, and a fuel tank.

Performance evaluation can be defined by the probability of detection (or detection rate), P_d, and the probability of false alarm, P_{fa}. The detection rate is determined by counting the total number of objects that are located at the surveyed positions of the emplaced targets and normalizing this total by the number of emplaced targets [296]. The probability of false alarm is defined as:

$$P_{fa} = N_{fa}/\{(A_S - A_T/\langle A_{fa}\rangle\}$$ (7.1)

where: N_{fa} = number of objects that are not associated with targets
 $\langle A_{fa}\rangle$ = average false alarm area

A_S = area of the site
A_T = area associated with targets

The ratio of the areas represents the opportunities for false alarms.

The main difficulty with the data collected from many of the landmine experiments is that the number of mine targets is small (e.g. 20 targets per scenario), and the clutter data is measured on small areas (e.g. 100–200 m^2) [298]. Since the two primary measures of detection performance, probability of detection and false alarm rate, are statistical measures, the large statistical uncertainties associated with measurements on small data sets impose a severe limitation on the interpretation and extrapolation of the test results. As a result a simple model has been suggested to estimate the uncertainty in a probability of detection measurement [298]. If detections can be described as a binomial process weighted by the true probability of detection, the uncertainty in the number of targets detected will be the square root of the number of targets detected, N, for large N. The percentage uncertainty in the probability of detection is:

$$U_{Pd}(\%) = \left[(\sqrt{N})/N\right] \times 100 \quad \text{(for large } N\text{)} \tag{7.2}$$

A similar statistical uncertainty is encountered in the measurement of probability of false alarm. An opportunity for a false alarm can be defined by the amount of ground covered by the projected area of the target plus an allowable miss distance, for example 1 m^2. Thus, for each 1 m^2 of ground that it passes over, the system has an opportunity of declaring a false alarm. A test field having an area of 20 m^2 offers 20 samples for probability of false alarm measurement. In a larger site there will be an increase in the clutter environment sampled by the sensor and an accompanying improvement in false alarm measurements.

References

1. Yinon, J. and Zitrin, S. *Modern Methods and Applications in Analysis of Explosives*, John Wiley & Sons, Chichester, 1993.
2. Gordon, L. and Hartley, W. R. 2,4,6-Trinitrotoluene (TNT). In: Roberts, W. C. and Hartley, W. R. (eds), *Drinking Water Health Advisory: Munitions*, Lewis Publishers, Boca Raton, FL, 1992, pp. 327–398.
3. Urbanski, T. *Chemistry and Technology of Explosives*, Vols 1–3, Pergamon Press, Oxford, 1964.
4. Urbanski, T. *Chemistry and Technology of Explosives*, Vol. 4, Pergamon Press, Oxford, 1984.
5. Yinon, J. and Zitrin, S. *The Analysis of Explosives*, Pergamon Press, Oxford, 1981.
6. *Kirk-Othmer Encyclopedia of Chemical Technology*, 3rd edn., Vol. 9, John Wiley & Sons, New York, 1980, pp. 561–620.
7. *Merck Index*, 9th edn., Merck & Co. Inc., New York, 1976.
8. Ro, K. S., Venugopal, A., Adrian, D. D., Constant, D., Qaisi, K., Valsaraj, K. T., Thibodeaux, L. J. and Roy, D. Solubility of 2,4,6-Trinitrotoluene in water. *Journal of Chemical & Engineering Data*, 1996, **41**, 758–761.
9. Alm, A., Dalman, O., Frolen-Lindgren, I., Hulten, F., Karlsson, T. and Kowalska, M. *Analyses of Explosives*, National Defence Research Institute, Stockholm, 1978.
10. Dionne, B. C., Rounbehler, D. P., Achter, E. K., Hobbs, J. R. and Fine, D. H. Vapor pressure of explosives. *Journal of Energetic Materials*, 1986, **4**, 447–472.
11. Yinon, J. *Toxicity and Metabolism of Explosives*, CRC Press, Boca Raton, FL, 1990.
12. McLellan, W. L., Hartley, W. R. and Brower, M. E. Hexahydro-1,3,5-trinitro-1,3,5-triazine (RDX). In: Roberts, W. C. and Hartley, W. R. (eds), *Drinking Water Health Advisory: Munitions*, Lewis Publishers, Boca Raton, FL, 1992, pp. 133–180.
13. Patterson, J., Brown, J., Duckert, W., Polson, J. and Shapira, N. I. *State-of-the-art: Military Explosives and Propellants Production Industry. Vol 1.* Report No. EPA-600/2-76-213a, American Defense Preparedness Association, Washington, DC, 1976.
14. McLellan, W. L., Hartley, W. R. and Brower, M. E. Octahydro-1,3,5,7-tetranitro-1,3,5,7-tetrazocine (HMX). In: Roberts, W. C. and Hartley, W. R. (eds), *Drinking Water Health Advisory: Munitions*, Lewis Publishers, Boca Raton, FL 1992, pp. 247–273.

15. McCrone, W. C. Cyclotetramethylene tetranitramine (HMX). *Analytical Chemistry*, 1950, **22**, 1225–1226.
16. Merril, E. J. Solubility of pentaerythritol tetranitrate-1,2-C^{14} in water and saline. *Journal of Pharmaceutical Science*, 1965, **54**, 1670–1671.
17. von Oettingen, W. F. and Donahue, D. D. *Toxicity and Potential Dangers of Pentaerythritol Tetranitrate (PETN). Acute Toxic Manifestations of PETN.* US Public Health Bulletin, Report No. 282, Washington, DC, 1944.
18. Hartley, W. R., Glennon, J., Gordon, L. and Normandy, J. Trinitroglycerol (TNG). In: Roberts, W. C. and Hartley, W. R. (eds), *Drinking Water Health Advisory: Munitions*, Lewis Publishers, Boca Raton, FL 1992, pp. 275–326.
19. Fedoroff, B. T., (ed.), *Encyclopedia of Explosives and Related Items*, Vols 1–10, Picatinny Arsenal, Dover, NJ, 1960–83.
20. Lund, R. P., Haggendal, J. and Johnsson, G. Withdrawal symptoms in workers exposed to nitroglycerine. *British Journal of Industrial Medicine*, 1968, **25**, 136–138.
21. Midkiff, J. C. R. and Walters, A. N. Slurry and emulsion explosives: new tools for terrorists, new challenges for detection and identification. In: Yinon J. (ed.), *Advances in Analysis and Detection of Explosives*, Kluwer Academic Publishers, Dordrecht, 1993, pp. 77–90.
22. Persson, P.-A., Holmberg, R. and Lee, J. *Rock Blasting and Explosives Engineering*, CRC Press, Boca Raton, FL, 1994, p. 540.
23. McGann, W., Jenkins, A. and Ribeiro, K. A thermodynamic study of the vapor pressures of C-4 and pure RDX. *First International Symposium on Explosive Detection Technology*, Atlantic City, NJ, 1991, pp. 518–531.
24. Fine, D. H., Rounbehler, D. P. and Curby, W. A. Dichotomous key approach for high confidence level identification of selected explosive vapors. *First International Symposium on Explosive Detection Technology*, Atlantic City, NJ, 1991, pp. 505–517.
25. Lovett, S. Explosive search dogs. *First International Symposium on Explosive Detection Technology*, Atlantic City, NJ, 1991, pp. 774–775.
26. Wright, R. H. *The Sense of Smell*, CRC Press, Boca Raton, 1982, p. 236.
27. Ternes, J. W. Integration of the human, canine, machine interface for explosives detection. *First International Symposium on Explosive Detection Technology*, Atlantic City, NJ, 1991, pp. 891–902.
28. Williams, M., Johnston, J. M., Waggoner, L. P., Jackson, J., Jones, M., Boussom, T., Hallowell, S. F. and Petrousky, J. A. Determination of the canine detection signature for NG smokeless powder. *Second Explosives Detection Technology Symposium & Aviation Security Technology Conference*, Atlantic City, NJ, 1996, pp. 328–339.
29. Luescher, U. A. Animal behavior case of the month. *Journal of the American Veterinary Medical Association*, 1993, **203**, 1538–1539.
30. Weinstein, S., Drozdenko, R. and Weinstein, C. Detection of extremely low concentrations of ultra pure TNT by rat. *Third International Symposium on Analysis and Detection of Explosives*, Mannheim-Neuostheim, Germany, 1989, pp. 32.31–32.34.
31. Biederman, G. B. Vapor preconcentration in the detection of explosives by animals in an automated setting. In: Yinon J. (ed.), *Advances in Analysis and Detection of Explosives*, Kluwer Academic Publishers, Dordrecht, 1993, pp. 463–472.
32. Peterson, P. K. Temperature dependence of adsorption effects of explosives molecules. *First International Symposium on Analysis and Detection of Explosives*, FBI Academy, Quantico, VA, 1983, pp. 391–395.

33. Conrad, F. J. Explosives detection. The problem and prospects. *Nuclear Materials Management*, 1984, **13**, 212–215.
34. Fraim, F. W., Achter, E. K., Carroll, A. L. and Hainsworth, E. Efficient collection of explosive vapors, particles and aerosols. *First International Symposium on Explosive Detection Technology*, Atlantic City, NJ, 1991, pp. 559–570.
35. Nacson, S. Adsorption phenomena in explosive detection. *Fifth International Symposium on Analysis and Detection of Explosives*, Washington, DC, 1995, paper no. 41.
36. Neudorfl, P., McCooeye, M. A. and Elias, L. Testing protocol for surface-sampling detectors. In: Yinon J. (ed.), *Advances in Analysis and Detection of Explosives*, Kluwer Academic Publishers, Dordrecht, 1993, pp. 373–384.
37. Davidson, W. R., Stott, W. R., Sleeman, R. and Akery, A. K. Synergy or dichotomy—vapor and particle sampling in the detection of contraband. *Conference on Substance Detection Systems*, Innsbruck, Austria, 1993. *SPIE Proceedings*, 1994, **2092**, 108–119.
38. Nacson, S., McNelles, L., Nargolwalla, S. and Greenberg, D. Method of detecting taggants in plastic explosives. Airport trials and solubility of explosives. *Second Explosives Detection Technology Symposium & Aviation Security Technology Conference*, Atlantic City, NJ, 1996, pp. 38–48.
39. Fine, D. H., Lieb, D. and Rufeh, F. Principle of operation of the thermal energy analyzer for the trace analysis of volatile and non-volatile *N*-nitroso compounds. *Journal of Chromatography*, 1975, **107**, 351–357.
40. Lafleur, A. L. and Mills, K. M. Trace level determination of selected nitroaromatic compounds by gas chromatography with pyrolysis/chemiluminescent detection. *Analytical Chemistry*, 1981, **53**, 1202–1205.
41. Goff, E. U., Yu, W. C. and Fine, D. H. Description of a nitro/nitroso specific detector for the analysis of explosives. *International Symposium on Analysis and Detection of Explosives*, Quantico, VA, 1983, pp. 159–168.
42. Rounbehler, D. P., MacDonald, S. J., Lieb, D. P. and Fine, D. H. Analysis of explosives using high speed gas chromatography with chemiluminescent detection. *First International Symposium on Explosive Detection Technology*, Atlantic City, NJ, 1991, pp. 703–713.
43. Wohltjen, H. and Dessy, R. Surface acoustic wave probe for chemical analysis. 1. Introduction and instrument description. *Analytical Chemistry*, 1979, **51**, 1458–1464.
44. Wohltjen, H. and Dessy, R. Surface acoustic wave probes for chemical analysis. 2. Gas chromatography detector. *Analytical Chemistry*, 1979, **51**, 1465–1470.
45. Watson, G., Horton, W. and Staples, E. Vapor detection using SAW sensors. *First International Symposium on Explosive Detection Technology*, Atlantic City, NJ, 1991, pp. 589–603.
46. Staples, E. J., Watson, G. W. A gas chromatograph incorporating an innovative new surface acoustic wave (SAW) detector. *Pittsburgh Conference on Analytical Chemistry and Applied Spectroscopy*, New Orleans, LA, 1998, paper no. 1583CP.
47. McDowell, C. A. *Mass Spectrometry*, McGraw Hill, New York, 1963.
48. Dawson, P. H. Principles of operation. In: Dowson P. H. (ed.), *Quadrupole Mass Spectrometry and its Applications*, Elsevier, Amsterdam, 1976, pp. 9–64.
49. Louris, J. N., Cooks, R. G., Syka, J. E. P., Kelley, P. E., Stafford, Jr., G. C. and Todd, J. F. J. Instrumentation, applications, and energy deposition in quadrupole ion-trap tandem mass spectrometry. *Analytical Chemistry*, 1987, **59**, 1677–1685.
50. Cooks, R. G., Glish, G. L., McLuckey, S. A. and Kaiser, R. E. Ion trap mass spectrometry. *Chemical & Engineering News*, 1991, pp. 26–41.
51. McLafferty, F. W. Tandem mass spectrometry. *Science*, 1981, **214**, 280–287.

52. Yinon, J. MS/MS techniques in forensic science. In: Maehly, A. and Williams, R. L. (eds), *Forensic Science Progress*, Vol 5, Springer-Verlag, Heidelberg, 1991, pp. 1–29.

53. Yost, R. A. and Enke, C. G. Triple quadrupole mass spectrometry for direct mixture analysis and structure elucidation. *Analytical Chemistry*, 1979, **51**, 1251A–1264A.

54. Quarmby, S. T. and Yost, R. A. *An improved method for performing a scan function on a quadrupole ion trap mass spectrometer.* US Patent No. 08/837,030. April 15, 1997.

55. Yinon, J., McClellan, J. E. and Yost, R. A. Electrospray ionization tandem mass spectrometry collision-induced dissociation study of explosives in an ion trap mass spectrometer. *Rapid Communications in Mass Spectrometry*, 1997, **11**, 1961–1970.

56. Davidson, W. R., Thomson, B. A., Akery, A. K. and Sleeman, R. Modifications to the ionization process to enhance the detection of explosives by API/MS/MS. *First International Symposium on Explosive Detection Technology*, Atlantic City, NJ, 1991, pp. 653–662.

57. Davidson, W. R., Stott, W. R., Akery, A. K. and Sleeman, R. The role of mass spectrometry in the detection of explosives. *First International Symposium on Explosive Detection Technology*, Atlantic City, NJ, 1991, pp. 663–671.

58. Sleeman, R., Bennett, G., Davidson, W. R. and Fisher, W. The detection of illicit drugs and explosives in real-time by tandem mass spectrometry. *International Symposium on Contraband and Cargo Inspection Technologies*, Washington, DC, 1992, pp. 57–63.

59. Stott, W. R., Davidson, W. R. and Sleeman, R. High throughput real time chemical contraband detection. *Conference on Substance Detection Systems*, Innsbruck, Austria, 1993. *SPIE Proceedings*, 1994, **2092**, 53–63.

60. Bennett, G., Sleeman, R., Davidson, W. R. and Stott, W. R. An airport trial of a system for the mass screening of baggage or cargo. *Conference on Cargo Inspection Technologies*, San Diego, CA, 1994. *SPIE Proceedings*, 1994, **2276**, 363–371.

61. McLuckey, S. A., Glish, G. L., Asano, K. G. and Grant, B. C. Atmospheric sampling glow discharge ionization source for the determination of trace organic compounds in ambient air. *Analytical Chemistry*, 1988, **60**, 2220–2227.

62. McLuckey, S. A., Glish, G. L. and Asano, K. G. Coupling of an atmospheric-sampling ion source with an ion-trap mass spectrometer. *Analytica Chimica Acta*, 1989, **225**, 25–35.

63. McLuckey, S. A., Goeringer, D. E., Asano, K. G., Vaidyanathan, G. and Stephenson, Jr., J. L. High explosives vapor detection by glow discharge-ion trap mass spectrometry. *Rapid Communications in Mass Spectrometry*, 1996, **10**, 287–298.

64. Lee, H. G., Lee, E. D. and Lee, M. L. Atmospheric pressure ionization time-of-flight mass spectrometer for real-time explosive vapor detection. *First International Symposium on Explosive Detection Technology*, Atlantic City, NJ, 1991, pp. 619–639.

65. West, R., Wu, M., Sin, J., Lee, W., Woolley, C., Fabbi, J. and Lee, M. L. Detection and identification of explosives using a TOFMS vapor analyzer. *Second Explosives Detection Technology Symposium & Aviation Security Technology Conference*, Atlantic City, NJ, 1996, pp. 100–105.

66. Boumsellek, S., Alajajian, S. H. and Chutjian, A. Negative-ion formation in the explosives RDX, PETN, and TNT by using the reversal electron attachment detection technique. *Journal of the American Society for Mass Spectrometry*, 1993, **3**, 243–247.

67. Chutjian, A. and Darrach, M. R. Improved, portable reversal electron attachment (READ) vapor detection system for explosives detection. *Second Explosives Detection Technology Symposium & Aviation Security Technology Conference*, Atlantic City, NJ, 1996, pp. 176–180.
68. Yinon, J., Boettger, H. G. and Weber, W. P. Negative-ion mass spectrometry. A new analytical method for the detection of TNT. *Analytical Chemistry*, 1972, **44**, 2235–2237.
69. Yinon, J. and Boettger, H. G. Modification of a high resolution mass spectrometer for negative ionization. *International Journal of Mass Spectrometry and Ion Physics*, 1972/73, **10**, 161–168.
70. Laramee, J. A., Kocher, C. A. and Deinzer, M. L. Application of a trochoidal electron monochromator/mass spectrometer system to the study of environmental chemicals. *Analytical Chemistry*, 1992, **64**, 2316–2322.
71. Laramee, J. A., Cody, R. B., Herrmannsfeldt, W. B. and Deinzer, M. L. Tunable energy electron monochromator interface to a sector instrument. *44th ASMS Conference on Mass Spectrometry and Allied Topics*, Portland, OR, 1996, p. 71.
72. Laramee, J. A. and Deinzer, M. L. Tunable energy (0.03–30 eV) electron capture negative ion mass spectrometry. *Fifth International Symposium on Analysis and Detection of Explosives*, Washington, DC, 1995, Paper No. 49.
73. Karpas, Z. Forensic science applications of ion mobility spectrometry. *Forensic Science Review*, 1989, **1**, 103–119.
74. St. Louis, R. H. and Hill, Jr., H. H. Ion Mobility Spectrometry in Analytical Chemistry. *Critical Reviews in Analytical Chemistry*, 1990, **21**, 321–355.
75. Fetterolf, D. D. and Clark, T. D. Detection of trace explosive evidence by ion mobility spectrometry. *Journal of Forensic Sciences*, 1993, **38**, 28–39.
76. Danylewych-May, L. L. and Cumming, C. Explosive and taggant detection with Ionscan. In: Yinon J. (ed.), *Advances in Analysis and Detection of Explosives*, Kluwer Academic Publishers, Dordrecht, 1993, pp. 385–401.
77. Anon, *Examining the analytical capabilities of a new hand portable gas chromatography/ion mobility spectrometry system*. http://femtoscan.com/analcap.htm, Application Note No. 9702, FemtoScan Corporation, 1998.
78. Henderson, D. O., Silberman, E. and Snyder, F. W. Fourier-transform infrared spectroscopy applied to explosive vapor detection. *First International Symposium on Explosive Detection Technology*, Atlantic City, NJ, 1991, pp. 604–617.
79. Janni, J., Gilbert, B. D., Field, R. W. and Steinfeld, J. I. Infrared absorption of explosive molecule vapors. *Spectrochimica Acta Part A*, 1997, **53**, 1375–1381.
80. Riris, H., Carlisle, C. B., McMillen, D. F. and Cooper, D. E. Explosives detection with a frequency modulation spectrometer. *Applied Optics*, 1996, **35**, 4694–4704.
81. Seitz, W. R. and Neary, M. P. Chemiluminescence and bioluminescence in chemical analysis. *Analytical Chemistry*, 1974, **46**, 188A–202A.
82. Boncyk, E. M. A bioluminescent explosives vapor detection and identification system. *Third International Symposium on Analysis and Detection of Explosives*, Mannheim-Neuostheim, Germany, 1989, pp. 40.41–40.14.
83. Rosengren, L.-G. Optimal optoacoustic detector design. *Applied Optics*, 1975, **14**, 1960–1976.
84. Claspy, P. C., Pao, Y.-H., Kwong, S. and Nodov, E. Laser optoacoustic detection of explosive vapors. *Applied Optics*, 1976, **15**, 1506–1509.
85. Claspy, P. C. Infrared optoacoustic spectroscopy and detection. In: Pao Y.-H. (ed.), *Optoacoustic Spectroscopy and Detection*, Academic Press, New York, 1977, pp. 133–166.
86. Crane, R. A. Laser optoacoustic absorption spectra for various explosive vapors. *Applied Optics*, 1978, **17**, 2097–2102.

87. Hobbs, J. R. and Conde, E. P. Analysis of airflows in personnel screening booths. In: Yinon J. (ed.), *Advances in Analysis and Detection of Explosives*, Kluwer Academic Publishers, Dordrecht, 1993, pp. 437–453.

88. Bromberg, E. E. A., Dussault, D., MacDonald, S. and Curby, W. A. Vapor generation for use in explosive portal detection devices. In: Yinon J. (ed.), *Advances in Analysis and Detection of Explosives*, Kluwer Academic Publishers, Dordrecht, 1993, pp. 473–484.

89. Hintze, M. M., Hansen, B. L. and Heath, R. L. Real-time explosives/narcotics vapor enhancement and collection systems for use with the atmospheric pressure ionization time-of-flight mass spectrometer. *First International Symposium on Explosive Detection Technology*, Atlantic City, NJ, 1991, pp. 634–636.

90. Jenkins, A., McGann, W. and Ribeiro, K. Extraction, transportation and processing of explosives vapor in detection systems. *First International Symposium on Explosive Detection Technology*, Atlantic City, NJ, 1991, pp. 532–551.

91. Wendel, G. J., Bromberg, E. E. A., Durfee, M. K. and Curby, W. Design of a walk-through trace EDS. *Second Explosives Detection Technology Symposium & Aviation Security Technology Conference*, Atlantic City, NJ, 1996, pp. 181–186.

92. Parmeter, J. E., Linker, K. L., Rhykerd, Jr., C. L. and Hannum, D. W. Testing of a walk-through portal for the trace detection of contraband explosives. *Second Explosives Detection Technology Symposium & Aviation Security Technology Conference*, Atlantic City, NJ, 1996, pp. 187–192.

93. Rhykerd, C., Linker, K., Hannum, D. and Parmeter, J. Walk-through explosives detection portal. *Nuclear Materials Management*, 1997, **26**, 97–102.

94. Lucero, D. P., Roder, S. R., Jankowski, P. and Mercado, A. Design concept: Femtogram level explosives vapor generator. In: Yinon J. (ed.), *Advances in Analysis and Detection of Explosives*, Kluwer Academic Publishers, Dordrecht, 1993, pp. 485–502.

95. Eiceman, G. A., Preston, D., Tiano, G., Rodriguez, J. and Parmeter, J. E. Quantitative calibration of vapor levels of TNT, RDX, and PETN using a diffusion generator with gravimetry and ion mobility spectrometry. *Talanta*, 1997, **45**, 57–74.

96. Reiner, G. A., Heisy, C. L. and McNair, H. M. A transient explosive vapor generator based on capillary gas chromatography. *Journal of Energetic Materials*, 1991, **9**, 173–190.

97. Davies, J. P., Blackwood, L. G., Davis, S. G., Goodrich, L. D. and Larsen, R. A. Design and calibration of pulsed vapor generators for TNT, RDX and PETN. In: Yinon J. (ed.), *Advances in Analysis and Detection of Explosives*, Kluwer Academic Publishers, Dordrecht, 1993, pp. 513–532.

98. Davies, J. P., Blackwood, L. G., Davis, S. G., Goodrich, L. D. and Larson, R. A. Design and calibration of pulsed vapor generators for 2,4,6-trinitrotoluene, cyclo-1,3,5-trimethylene-2,4,6-trinitramine, and pentaerythritol tetranitrate. *Analytical Chemistry*, 1993, **65**, 3004–3009.

99. Stott, W. R., Green, D. and Mercado, A. Mass spectrometric detection of solid and vapor explosive materials. *Conference on Cargo Inspection Technologies*, San Diego, CA, 1994. SPIE Proceedings, 1994, **2276**, 87–97.

100. Hartell, M. G., Myers, L. J., Wagonner, P., Kuhlman, M., Hallowell, S. F. and Petrousky, J. A. Design and testing of a quantitative vapor delivery system. *Fifth International Symposium on Analysis and Detection of Explosives*, Washington, DC, 1995, Paper No. 48.

101. Hartell, M. G., Pierce, M. Q., Myers, L. J., Hallowell, S. F. and Petrousky, J. A. Comparative analysis of smokeless powder vapor signatures derived under static versus dynamic conditions. *Fifth International Symposium on Analysis and Detection of Explosives*, Washington, DC, 1995, Paper No. 46.

102. Grodzins, L. Photons in - photons out: Non-destructive inspection of containers using X-ray and gamma ray techniques. *First International Symposium on Explosive Detection Technology*, Atlantic City, NJ, 1991, pp. 201–231.
103. Grodzins, L. Nuclear techniques for finding chemical explosives in airport luggage. *Nuclear Instruments and Methods in Physics Research B*, 1991, **56/57**, 829–833.
104. Anon. *Technology against terrorism: The federal effort*. US Congress, Office of Technology Assessment, Report No. OTA-ISC-481, Washington, DC, 1991.
105. Anon. *Technology against terrorism: Structuring security*. US Congress, Office of Technology Assessment, Report No. OTA-ISC-511, Washington, DC, 1992.
106. Condon, E. U. X-rays. In: Condon E. U. and Odishaw E. U. (eds.), *Handbook of Physics*, McGraw-Hill, New York, 1967, pp. 7.126–127.138.
107. Schafer, D., Annis, M. and Hacker, M. New X-ray technology for the detection of explosives. *First International Symposium on Explosive Detection Technology*, Atlantic City, NJ, 1991, pp. 269–281.
108. Aitkenhead, W. F. and Stillson, J. H. Dual energy X-ray Compton scattering measurement as a method to detect sheet explosives. *Second Explosives Detection Technology Symposium & Aviation Security Technology Conference*, Atlantic City, NJ, 1996, pp. 236–241.
109. Luggar, R. D., Horrocks, J. A., Speller, R. D. and Lacey, R. J. Low angle X-ray scatter for explosives detection: a geometry optimization. *Applied Radiation and Isotopes*, 1997, **48**, 215–224.
110. Carter, T., Dermody, G., Pleasants, I. B., Burrows, D., Mackenzie, S. J., Jupp, I. D. and Ramsden, D. Angular resolved X-ray diffraction measurements of explosives. *Sixth International Symposium on Analysis and Detection of Explosives*, Prague, Czech Republic, 1998.
111. Hnatnicky, S. CXRS explosives detection airport prototype. *Conference on Physics-based Technologies for The Detection of Contraband*, Boston, MA, 1996. *SPIE Proc*, 1997, **2936**, 180–190.
112. Harding, G., Newton, M. and Kosanetzky, J. Energy-dispersive X-ray diffraction tomography. *Physics in Medicine and Biology*, 1990, **35**, 33–41.
113. Curry, I. T. S., Dowdey, J. E. and Murry, Jr., R. C. *Christensen's Physics of Diagnostic Radiology*, 4th ed., Lea & Febiger, Philadelphia, PA, 1990, p. 522.
114. Anon. Product overview. http://www.invision-tech.com/products/prodover.htm: In Vision Technologies, Newark, CA, 1998.
115. Dolan, K. W., Ryon, R. W., Schneberk, D. J., Martz, H. E. and Rikard, R. D. Explosives detection limitations using dual-energy radiography and computed tomography. *First International Symposium on Explosive Detection Technology*, Atlantic City, NJ, 1991, pp. 252–260.
116. Lu, Q. and Conners, R. W. X-ray image analysis for luggage detection. *Second Explosives Detection Technology Symposium & Aviation Security Technology Conference*, Atlantic City, NJ, 1996, pp. 242–247.
117. Willard, H. H., Merritt, Jr., L. L. and Dean, J. A. *Instrumental Methods of Analysis*, Van Nostrand Co., Princeton, NJ, 1961, p. 626.
118. Clifford, J. R., Miller, R. B., McCullough, W. F. and Habiger, K. W. An explosive detection system for screening luggage with high energy X-rays. *First International Symposium on Explosive Detection Technology*, Atlantic City, NJ, 1991, pp. 237–251.
119. Vartsky, D., Engler, G. and Goldberg, M. B. A method for detection of explosives based on nuclear resonance absorption of gamma rays in ^{14}N. *Nuclear Instruments and Methods in Physics Research A*, 1994, **348**, 688–691.
120. Sredniawski, J. J. Detecting concealed explosives with gamma rays. *The Industrial Physicist*, 1997, **3**, 24–27.

121. Rabenstein, D. L. Nuclear magnetic resonance spectroscopy. In: Gouw T. H. (ed.), *Guide to Modern Methods of Instrumental Analysis*, Wiley-Interscience, New York, 1972, pp. 231–278.

122. Hore, P. J. *Nuclear Magnetic Resonance*, Oxford University Press, Oxford, 1995.

123. McKay, D. R., Moeller, C. R., Magnuson, E. E. and Burnett, L. J. Liquid explosives screening system. *First International Symposium on Explosive Detection Technology*, Atlantic City, NJ, 1991, pp. 454–464.

124. Jelinski, L. W. Modern NMR spectroscopy. *Chemical & Engineering News, 5 November* 1984, 26–47.

125. King, J. D., De Los Santos, A., Nicholls, C. I. and Rollwitz, W. L. Application of magnetic resonance to explosives detection. *First International Symposium on Explosive Detection Technology*, Atlantic City, NJ, 1991, pp. 478–485.

126. King, J. D. Explosives detection—The case for magnetic resonance. In: Yinon J. (ed.), *Advances in Analysis and Detection of Explosives*, Kluwer Academic Publishers, Dordrecht, 1993, pp. 351–359.

127. Poole, Jr., C. P. Electron spin resonance. In: Gouw T. H. (ed.), *Guide to Modern Methods of Instrumental Analysis*, Wiley-Interscience, New York, 1972, pp. 279–321.

128. Poindexter, E. H., Leupold, H. A., Wittstruck, R. H. and Gerardi, G. J. Consideration of untried magnetic resonance techniques for possible baggage analysis. *First International Symposium on Explosive Detection Technology*, Atlantic City, NJ, 1991, pp. 493–504.

129. Laszlo, P. and Stang, P. J. *Organic Spectroscopy. Principles and Applications*, Harper & Row Publishers, New York, 1971.

130. Sanders, J. K. M. and Hunter, B. K. *Modern NMR Spectroscopy. A Guide for Chemists*, Oxford University Press, Oxford, 1988.

131. Garroway, A. N., Buess, M. L., Miller, J. B., McGrath, K. J., Yesinowski, J. P., Suits, B. H. and Miller, G. R. Nuclear Quadrupole resonance (NQR) for detection of explosives and landmines. *Sixth International Symposium on Analysis and Detection of Explosives*, Prague, Czech Republic, 1998.

132. Buess, M. L., Garroway, A. N., Miller, J. B. and Yesinowski, J. P. Explosives detection by ^{14}N pure NQR. In: Yinon J. (ed.), *Advances in Analysis and Detection of Explosives*, Kluwer Academic Publishers, Dordrecht, 1993, pp. 361–368.

133. Garroway, A. N., Buess, M. L., Yesinowski, J. P. and Miller, J. B. Narcotics and explosives detection by ^{14}N pure NQR. *Conference on Substance Detection Systems*, Innsbruck, Austria, 1993. *SPIE Proceedings*, 1994, **2092**, 318–327.

134. Buess, M. L., Garroway, A. N. and Miller, J. B. NQR detection using a meanderline surface coil, *Journal of Magnetic Resonance*, 1991, **92**, 348–362.

135. Rayner, T., Thorson, B., Beevor, S., West, R. and Krauss, R. Explosives detection using quadrupole resonance analysis. *Second Explosives Detection Technology Symposium & Aviation Security Technology Conference*, Atlantic City, NJ, 1996, pp. 275–280.

136. Rayner, T., Burnett, L. and West, R. Performance trade-offs in quadrupole resonance analysis screening. *Second Explosives Detection Technology Symposium & Aviation Security Technology Conference*, Atlantic City, NJ, 1996, pp. 287–292.

137. Krauss, R. A. Development testing of a quadrupole resonance explosives detection device. *Second Explosives Detection Technology Symposium & Aviation Security Technology Conference*, Atlantic City, NJ, 1996, pp. 281–286.

138. Grechishkin, V. S. NQR device for detecting plastic explosives, mines and drugs. *Applied Physics A*, 1992, **55**, 505–507.

139. Rudakov, T. N., Mikhaltsevich, V. T. and Selchikhin, O. P. The use of multipulse nuclear quadrupole resonance techniques for the detection of explosives containing RDX. *Journal of Physics D: Applied Physics*, 1997, **30**, 1377–1382.

140. Rayner, T. J., Hibbs, A. and Burnett, L. J. Quadrupole resonance explosive detection systems. *Sixth International Symposium on Analysis and Detection of Explosives*, Prague, Czech Republic, 1998.

141. Falconer, D. G. and Watters, D. G. Explosive detection using microwave imaging. *First International Symposium on Explosive Detection Technology*, Atlantic City, NJ, 1991, pp. 486–492.

142. Sheen, D. M., McMakin, D. L., Collins, H. D., Hall, T. E. and Severtsen, R. H. Concealed explosive detection on personnel using a wideband holographic millimeter-wave imaging system. *Conference on Signal Processing, Sensor Fusion, and Target Recognition V*, Orlando, FL, 1996. *SPIE Proceedings*, 1996, **2755**, 503–513.

143. Seward, D. C. and Yukl, T. Explosive detection using dielectrometry. *First International Symposium on Explosive Detection Technology*, Atlantic City, NJ, 1991, pp. 441–453.

144. Seward, C. and Yukl, T. Dielectric portal for screening people. *Second Explosives Detection Technology Symposium & Aviation Security Technology Conference*, Atlantic City, NJ, 1996, pp. 162–169.

145. Colthup, N. B. Infrared and Raman spectroscopy. In: Gouw T. H. (ed.), *Guide to Modern Methods of Instrumental Analysis*, Wiley-Interscience, New York, 1972, pp. 195–229.

146. Lewis, I. R., Daniel, Jr., N. W., Chaffin, N. C., Griffiths, P. R. and Tungol, M. W. Raman spectroscopic studies of explosive materials: towards a fieldable explosives detector. *Spectrochimica Acta Part A*, 1995, **51**, 1985–2000.

147. Cheng, C., Kirkbride, T. E., Batchelder, D. N., Lacey, R. J. and Sheldon, T. G. In situ detection and identification of trace explosives by Raman microscopy. *Journal of Forensic Sciences*, 1995, **40**, 31–37.

148. Hayward, I. P., Kirkbride, T. E., Batchelder, D. N. and Lacey, R. J. Use of fiber optic probe for the detection and identification of explosive materials by Raman spectroscopy. *Journal of Forensic Sciences*, 1995, **40**, 883–884.

149. Hussein, E. M. A. Detection of explosive materials using nuclear radiation: a critical review. *Conference on Aviation Security Problems and Related Technologies: Critical Reviews*, San Diego, CA, 1992, pp. 126–136.

150. Gozani, T. and Shea, P. M. Nuclear based explosive detection systems-1992 status. In: Yinon J. (ed.) *Advances in Analysis and Detection of Explosives*, Kluwer Academic Publishers, Dordrecht, 1993, pp. 335–349.

151. Gozani, T., Morgado, R. E. and Seher, C. C. Nuclear-based techniques for explosive detection. *Journal of Energetic Materials*, 1986, **4**, 377–414.

152. Gozani, T., Seher, C. C. and Morgado, R. E. Nuclear-based techniques for explosive detection. *Third International Symposium on Analysis and Detection of Explosives*, Mannheim-Neuostheim, Germany, 1989, pp. 36.31–36.19.

153. Shea, P., Gozani, T. and Bozorgmanesh, H. A TNA explosives-detection system in airline baggage. *Nuclear Instruments and Methods in Physics Research A*, 1990, **299**, 444–448.

154. Seher, C. C. Airport tests of federal aviation administration thermal neutron activation explosive detection systems. *Third International Symposium on Analysis and Detection of Explosives*, Mannheim-Neuostheim, Germany, 1989, pp. 34.31–34.11.

155. Caffrey, A. J., Cole, J. D., Gehrke, R. J. and Greenwood, R. C. Chemical warfare agent and high explosive identification by spectroscopy of neutron-induced gamma rays. *IEEE Transactions in Nuclear Science*, 1992, **39**, 1422–1426.

156. Gozani, T. Novel applications of fast neutron interrogation methods. *Nuclear Instruments and Methods in Physics Research A*, 1994, **353**, 635–640.

157. Vourvopoulos, G. and Schultz, F. J. A pulsed fast-thermal neutron system for the detection of hidden explosives. *Nuclear Instruments and Methods in Physics Research B*, 1993, **79**, 585–588.
158. Sawa, Z. P. PFN GASCA technique for detection of explosives and drugs. *Nuclear Instruments and Methods in Physics Research B*, 1993, **79**, 593–596.
159. Gozani, T. Understanding the physics limitations of PFNA - the nanosecond pulsed fast neutron analysis. *Nuclear Instruments and Methods in Physics Research B*, 1995, **99**, 743–747.
160. Stevenson, J., Clayton, J., Fankhauser, K., Gozani, T., Jeppesen, R., Liu, F., Merics, T. and Bell, C. PFNA detection of small explosives. *13th International Conference on Applications of Accelerators in Research and Industry*, Denton, TX, 1994.
161. Fink, C. L., Micklich, B. J., Yule, T. J., Humm, P., Sagalovsky, L. and Martin, M. M. Evaluation of neutron techniques for illicit substance detection. *Nuclear Instruments and Methods in Physics Research B*, 1995, **99**, 748–752.
162. Lefevre, H. W., Rasmussen, R. J., Chmelik, M. S., Schofield, R. M. S., Sieger, G. E. and Overley, J. C. Using a fast-neutron spectrometer system to candle luggage for hidden explosives. *International Conference on Neutrons in Research and Industry*, Crete, Greece, 1996. *SPIE Proceedings*, 1997, **2867**, 206–210.
163. Loveman, R. A., Feinstein, R. L., Bendahan, J., Gozani, T. and Shea, P. Laminography using resonant neutron attenuation for detection of drugs and explosives. In: Duggan J. L. and Morgan, I. L. (eds.), *Application of Accelerators in Research and Industry*, Vol. 392, American Institute of Physics, Denton, TX, 1997, pp. 895–898.
164. Gokhale, P. P. and Hussein, E. M. A. A ^{252}Cf neutron transmission technique for bulk detection of explosives. *Applied Radiation and Isotopes*, 1997, **48**, 973–979.
165. Rasmussen, R. J., Fanselow, W. S., Lefevre, H. W., Chmelik, M. S., Overley, J. C., Brown, A. P., Sieger, G. E. and Schofield, R. M. S. Average atomic number of heterogeneous mixtures from the ratio of gamma to fast-neutron attenuation. *Nuclear Instruments and Methods in Physics Research B*, 1997, **124**, 611–614.
166. Gomberg, H. J. and Kushner, B. G. Neutron elastic scatter (NES) for explosives detection systems (EDS). *First International Symposium on Explosive Detection Technology*, Atlantic City, NJ, 1991, pp. 123–139.
167. Beyerle, A., Hurley, J. P. and Tunnell, L. Design of an associated particle imaging system. *Nuclear Instruments and Methods in Physics Research A*, 1990, **299**, 458–462.
168. Beyerle, A., Durkee, R., Headley, G., Hurley, J. P. and Tunnell, L. Associated particle imaging. *First International Symposium on Explosive Detection Technology*, Atlantic City, NJ, 1991, pp. 160–174.
169. Lee, W., Mahood, D., Ryge, P., Shea, P. and Gozani, T. Thermal neutron analysis (TNA) explosive detection based on electronic neutron generators. *Nuclear Instruments and Methods in Physics Research B*, 1995, **99**, 739–742.
170. Fainberg, A. Explosives detection for aviation security. *Science*, 1992, **255**, 1531–1537.
171. Anon. *Detection of Explosives for Commercial Aviation Security*. National Research Council, Commission on Engineering and Technical Systems, National Materials Advisory Board, Report No. NMAB-71, Washington, DC, 1993.
172. Leung, V., Shea, P., Liu, F. and Sivakumar, M. Fusion of the nuclear and different X-ray technologies for explosives detection. *First International Symposium on Explosive Detection Technology*, Atlantic City, NJ, 1991, pp. 311–315.
173. Pongratz, H. W. Explosive detection for checked luggage by DETEX, a combined X-ray and positron-tomograph system. *First International Symposium on Explosive Detection Technology*, Atlantic City, NJ, 1991, pp. 319–332.

174. Anon. *Insight Eagle.* http://www.biosterile.com/insight_eagle.htm, BioSterile Technology, Inc., Ft. Wayne, IN, 1998.
175. Smith, M. C. and Hoopengardner, R. L. A systems approach to the explosive detection problem. *First International Symposium on Explosive Detection Technology,* Atlantic City, NJ, 1991, pp. 880–890.
176. Sheldon, T. G., Lacey, R. J., Murray, N. C. and Smith, G. M. Performance testing of explosives and weapons detection systems. *First International Symposium on Explosive Detection Technology,* Atlantic City, NJ, 1991, pp. 368–375.
177. Navarro, J. A., Becker, D. A., Kenna, B. T. and Kossack, C. F. A general protocol for operational testing and evaluation of bulk explosive detection systems. *First International Symposium on Explosive Detection Technology,* Atlantic City, NJ, 1991, pp. 347–367.
178. Elias, L. Development of trace explosive detection standards. *Journal of Testing and Evaluation,* 1994, **22**, 280–281.
179. Scharer, J. ICAO standard suitcase and standard box. *Third Meeting of ICAO Ad Hoc Group of Experts on Detection of Explosives,* Montreal, 1990, pp. AH-DE/3–14.
180. Boyars, C., Combs, D., Copeland, R., Fuller, G., Moler, R., Pate, C., Roberts, G., Roder, F. and Williams, P. *Explosives Tagging and Control Annual Report.* The Aerospace Corporation, Washington, DC, 1977.
181. Achter, E., Boyars, C., Combs, D., Copeland, J. R., Edwards, D., Fuller, G., Lucero, D., Markham, D., Moler, R., Pate, C., Roberts, G., Roder, F. and Sahmel, R. *Explosives Tagging and Control Second Annual Report.* The Aerospace Corporation, Washington, DC, 1978.
182. Peterson, A. A. A report on the detection and identification of explosives by tagging. *Journal of Forensic Sciences,* 1981, **26**, 313–318.
183. Scharer, J. The tagging of explosives. The new Swiss law on explosives: development, achievements and first experiences. *First International Symposium on Analysis and Detection of Explosives,* Quantico, VA, 1983, pp. 463–470.
184. Eng, S., Lannon, J. and Westerdahl, C. Program to tag military plastic/sheet explosives. *First International Symposium on Explosive Detection Technology,* Atlantic City, NJ, 1991, pp. 734–736.
185. Anon. *Containing the threat from illegal bombings. An integrated national strategy for marking, tagging, rendering inert, and licensing explosives and their precursors.* National Research Council, Washington, DC, 1998.
186. Kolla, P. Marking of German plastic explosive for the enhancement of vapor detection. *First International Symposium on Explosive Detection Technology,* Atlantic City, NJ, 1991, pp. 723–733.
187. Chen, T. H., Campell, C., Reed, R., Ark, W. F., Autera, J., Wiegand, D. A. and Kirshenbaum, M. S. Plant laboratory scale preparation and complete characterization of modified composition C-4 containing a taggant. *First International Symposium on Explosive Detection Technology,* Atlantic City, NJ, 1991, pp. 739–740.
188. Reed, R. A., Campell, C. and Chen, T. H. Prediction of the life-time of a taggant in a composition. In: Yinon J. (ed.), *Advances in Analysis and Detection of Explosives,* Kluwer Academic Publishers, Dordrecht, 1993, pp. 403–408.
189. Mostak, P., Stancl, M. and Preussler, V. Consideration of some aspects of marking plastic explosive Semtex. In: Yinon J. (ed.), *Advances in Analysis and Detection of Explosives,* Kluwer Academic Publishers, Dordrecht, 1993, pp. 429–436.
190. Mostak, P. and Stancl, M. Marking of emulsion explosives for detection. *Fifth International Symposium on Analysis and Detection of Explosives,* Washington, DC, 1995, Paper No. 55.

191. Jones, D. E. G., Augsten, R. A. and Feng, K. K. Detection agents for explosives. *Journal of Thermal Analysis*, 1995, **44**, 533–546.

192. Jones, D. E. G., Augsten, R. A., Murnaghan, K. P., Handa, Y. P. and Ratcliffe, C. I. Characterization of DMNB, a detection agent for explosives, by thermal analysis and solid state NMR, *Journal of Thermal Analysis*, 1995, **44**, 547–561.

193. Jones, D. E. G. and Feng, H. T. A preliminary study of the thermal decomposition of DMNB, an explosive detection agent, using a heat flux calorimeter. *International Autumn Seminar on Propellants, Explosives and Pyrotechnics*, Shenzhen, China, 1997.

194. Mostak, P. and Stancl, M. Marking of explosives. *Sixth International Symposium on Analysis and Detection of Explosives*, Prague, Czech Republic, 1998.

195. Mintz, K. J., Fouchard, R. and Elias, L. Microencapsulation of the marking agent 2,3-dimethyl-2,3-dinitrobutane. *Sixth International Symposium on Analysis and Detection of Explosives*, Prague, Czech Republic, 1998.

196. Mintz, K. J. and Elias, L. Marking of detonating cord using 2,3-dimethyl-2,3-dinitrobutane (DMNB). *Sixth International Symposium on Analysis and Detection of Explosives*, Prague, Czech Republic, 1998.

197. Porter, S. J. *Method of desensitizing fertilizer grade ammonium nitrate and the product obtained*. US Patent No. 3,366,468. 1968.

198. Nacson, S., Mitchner, B., Legrady, O., Siu, T. and Nargolwalla, S. A GC/ECD approach for the detection of explosives and taggants. *First International Symposium on Explosives Detection Technology*, Atlantic City, NJ, 1991, pp. 714–721.

199. Chen, T. H., Campell, C. and Reed, R. A. Detection of taggant in modified composition C-4 tagged with a taggant. *First International Symposium on Explosive Detection Technology*, Atlantic City, NJ, 1991, pp. 741–751.

200. Williams, D., Staples, E. J., Watson, G. W. and McGuire, D. A new surface acoustic wave detector and analysis of plastic explosives and nitro-taggants. *Second Explosives Detection Technology Symposium & Aviation Security Technology Conference*, Atlantic City, NJ, 1996, pp. 106–112.

201. Staples, E. J., Watson, G. W., McGuire, D. and Williams, D. SAW/GC detection of taggants and other volatile compounds associated with contraband materials. *Conference on Chemistry- and Biology-based Technologies for the Detection of Contraband*, Boston, MA, 1996. *SPIE Proc*, 1997, **2937**, 57–65.

202. Burlinson, N. E., Kaplan, L. A. and Adams, C. E. *Photochemistry of TNT: investigation of the 'pink water' problem*. Report No. NOLTR73–172 (AD-76970), Naval Ordnance Laboratory, White Oak, Silver Spring, MD, 1973.

203. McCormick, N. G., Feeherry, F. E. and Levinson, H. S. Microbial transformation of 2,4,6-trinitrotoluene and other nitroaromatic compounds. *Applied and Environmental Microbiology*, 1976, **31**, 949–958.

204. Kaplan, D. L. and Kaplan, A. M. Thermophilic biotransformation of 2,4,6-trinitrotoluene under simulated composting conditions. *Applied and Environmental Microbiology*, 1982, **44**, 757–760.

205. Tamiri, T. and Zitrin, S. Capillary column gas chromatography/mass spectrometry of explosives. *Journal of Energetic Materials*, 1986, **4**, 215–237.

206. Farey, M. G. and Wilson, S. E. Quantitative determination of tetryl and its degradation products by high-pressure liquid chromatography. *Journal of Chromatography*, 1975, **114**, 261–265.

207. Walsh, M. E. *Environmental transformation products of nitroaromatics and nitramines*. Special Report No. 90-2 (ADA220-610), U.S. Army Cold Regions Research and Engineering Laboratory, Hanover, NH, 1990.

208. Yinon, J., Zitrin, S. and Tamiri, T. Reactions in the mass spectrometry of 2,4,6-*N*-tetranitro-*N*-methylaniline (tetryl). *Rapid Communications in Mass Spectrometry*, 1993, **7**, 1051–1054.

209. Kayser, E. G., Burlinson, N. E. and Rosenblatt, D. H. *Kinetics of hydrolysis and products of hydrolysis and photolysis of tetryl*. Technical Report No. 84-78 (ADA 153144), Naval Surface Weapons Center, Silver Spring, MD, 1984.

210. Yasuda, S. K. Separation and identification of tetryl and related compounds by two-dimensional thin-layer chromatography. *Journal of Chromatography*, 1970, **50**, 453–457.

211. McCormick, N. G., Cornell, J. H. and Kaplan, A. M. Biodegradation of hexahydro-1,3,5-trinitro-1,3,5-triazine. *Applied and Environmental Microbiology*, 1981, **42**, 817–823.

212. Patterson, J., Shapira, N. I., Brown, J., Duckert, W. and Polson, J. *State-of-the-art: Military explosives and propellants production industry. Vol 2*. Report No. 600/2-76-213b, American Defense Preparedness Association, Washington, DC, 1976.

213. Wendt, T. M., Cornell, J. H. and Kaplan, A. M. Microbial degradation of glycerol nitrates. *Applied and Environmental Microbiology*, 1978, **36**, 693–699.

214. Basch, A., Margalit, Y., Abramovich-Bar, S., Bamberger, Y., Daphna, D., Tamiri, T. and Zitrin, S. Decomposition products of PETN in post explosion analysis. *Journal of Energetic Materials*, 1986, **4**, 77–91.

215. Heller, C. A., Greni, S. R. and Erickson, E. D. Field detection of 2,4,6-trinitrotoluene in water by ion-exchange resins. *Analytical Chemistry*, 1982, **54**, 286–289.

216. Zhang, Y., Seitz, W. R. and Grant, C. L. A clear, amine-containing poly(vinyl chloride) membrane for in situ optical detection of 2,4,6-trinitrotoluene. *Analytica Chimica Acta*, 1989, **217**, 217–227.

217. Zhang, Y., Seitz, W. R., Sundberg, D. C. and Grant, C. L. *Preliminary development of a fiber optic sensor for TNT*. Special Report No. 88-4, U.S. Army Cold Regions Research and Engineering Laboratory, Hanover, NH, 1988.

218. Zhang, Y. and Seitz, W. R. Single fiber absorption measurements for remote detection of 2,4,6-trinitrotoluene. *Analytica Chimica Acta*, 1989, **221**, 1–9.

219. Arbuthnot, D., Bartholomew, D. U., Carr, R., Elkind, J. L., Gheorghiu, L., Melendez, J. L. and Seitz, W. R. Detection of a polynitroaromatic compound using a novel polymer-based multiplate sensor. *Conference on Detection and Remediation Technologies for Mines and Minelike Targets III*, Orlando, FL, 1998. *SPIE Proceedings*, 1998, **3392**, 432–440.

220. Jian, C. and Seitz, W. R. Membrane for in situ optical detection of organic nitro compounds based on fluorescence quenching. *Analytica Chimica Acta*, 1990, **237**, 265–271.

221. Stevanovic, S. and Mitrovic, M. Colorimetric method for semiquantitative determination of nitroorganics in water. *International Journal of Environmental Analytical Chemistry*, 1990, **40**, 69–76.

222. Medary, R. T. Inexpensive, rapid field screening test for 2,4,6-trinitrotoluene in soil. *Analytica Chimica Acta*, 1992, **258**, 341–346.

223. Jenkins, T. F. and Walsh, M. E. Development of field screening methods for TNT, 2,4-DNT and RDX in soil. *Talanta*, 1992, **39**, 419–428.

224. Van Emon, J. M. and Lopez-Avila, V. Immunochemical methods for environmental analysis. *Analytical Chemistry*, 1992, **64**, 78A–88A.

225. Eck, D. L., Kurth, M. J. and Macmillan, C. Trinitrotoluene and other nitroaromatic compounds—Immunoassay methods. In: Van Emon J. M. and Mumma R. O. (eds.), *Immunochemical Methods for Environmental Analysis*, ACS Symposium Series, Vol. 442, Chapter 9, American Chemical Society, Washington, DC, 1990, pp. 79–94.

226. Keuchel, C., Weil, L. and Niessner, R. Enzyme-linked immunosorbant assay for the determination of 2,4,6-trinitrotoluene and related nitroaromatic compounds. *Analytical Sciences*, 1992, **8**, 9–12.

227. Keuchel, C., Weil, L. and Niessner, R. Effect of the variation of the length of the spacer in a competitive enzyme immunoassay (ELISA) for the determination of 2,4,6-trinitrotoluene (TNT). *Fresenius Journal of Analytical Chemistry*, 1992, **343**, 143–143.

228. Keuchel, C. and Niessner, R. Rapid field screening test for determination of 2,4,6-trinitrotoluene in water and soil with immunofiltration. *Fresenius Journal of Analytical Chemistry*, 1994, **350**, 538–543.

229. Anon. *TNT RaPID Assay*, Product file, Strategic Diagnostics, Inc., Newark, DE, 1995.

230. Shriver-Lake, L. C., Breslin, K. A., Charles, P. T., Conrad, D. W., Golden, J. P. and Ligler, F. S. Detection of TNT in water using an evanescent wave fiber-optic biosensor. *Analytical Chemistry*, 1995, **67**, 2431–2435.

231. Bhatia, S. K., Shriver-Lake, L. C., Prior, K. J., Georger, J. H., Calvert, J. M., Bredehorst, R. and Ligler, F. S. Use of thiol-terminal silanes and heterobifunctional crosslinkers for immobilization of antibodies on silica surfaces. *Analytical Biochemistry*, 1989, **178**, 408–413.

232. Shriver-Lake, L. C., Donner, B. L. and Ligler, F. S. On-site detection of TNT with a portable fiber optic biosensor. *Environmental Science & Technology*, 1997, **31**, 837–841.

233. Whelan, J. P., Kusterbeck, A. W., Wemhoff, G. A., Bredehorst, R. and Ligler, F. S. Continuous-flow immunosensor for detection of explosives. *Analytical Chemistry*, 1993, **65**, 3561–3565.

234. Bart, J. C., Judd, L. L., Hoffman, K. E., Wilkins, A. M., Charles, P. T. and Kusterbeck, A. W. Detection and quantitation of the explosives TNT and RDX in groundwater using a continuous flow immunosensor. In: Aga D. S. and Thurman E. M. (eds), *Development and Applications of Immunoassays for Environmental Analysis*, ACS symposium Series, Vol. 657, American Chemical Society, Washington. D., 1996, pp. 210–220.

235. Bart, J. C., Judd, L. L., Hoffman, K. E., Wilkins, A. M. and Kusterbeck, A. W. Application of a portable immunosensor to detect the explosives TNT and RDX in groundwater samples. *Environmental Science & Technology*, 1997, **31**, 1505–1511.

236. Bart, J. C., Judd, L. L. and Kusterbeck, A. W., Environmental immunoassay for the explosive RDX using a fluorescent dye-labeled antigen and the continuous-flow immunosensor. *Sensors and Actuators B*, 1997, **38–39**, 411–418.

237. Narang, U., Gauger, P. R. and Ligler, F. S. Capillary-based displacement flow immunosensor. *Analytical Chemistry*, 1997, **69**, 1961–1964.

238. Narang, U., Gauger, P. R. and Ligler, F. S. A displacement flow immunosensor for explosive detection using microcapillaries. *Analytical Chemistry*, 1997, **69**, 2779–2785.

239. Bard, A. J. and Faulkner, L. R. *Electrochemical Methods. Fundamentals and Applications*, John Wiley & Sons, New York, 1980.

240. Cao, Z., Buttner, W. J. and Stetter, J. R. The properties and applications of amperometric gas sensors. *Electroanalysis*, 1992, **4**, 253–266.

241. Buttner, W. J., Findlay, M., Vickers, W., Davis, W. M., Cespedes, E. R., Cooper, S. and Adams, J. W. In situ detection of trinitrotoluene and other nitrated explosives in soils. *Analytica Chimica Acta*, 1997, **341**, 63–71.

242. Yinon, J., Yost, R. A. and Bulusu, S. Thermal decomposition characterization of explosives by pyrolysis-gas chromatography-mass spectrometry. *Journal of Chromatography A*, 1994, **688**, 231–242.

243. Wormhoudt, J., Shorter, J. H. and Kolb, C. E. Mechanisms underlying the cone penetrometer detection of energetic materials in soils. *Conference on Field Analytical Methods for Hazardous Wastes and Toxic Chemicals*, Las Vegas, NV, 1997.
244. Wormhoudt, J., Shorter, J. H., McManus, J. B., Kebabian, P. L., Zahniser, M. S., Davis, W. M., Cespedes, E. R. and Kolb, C. E. Tunable infrared laser detection of pyrolysis products of explosives in soils. *Applied Optics*, 1996, **35**, 3992–3997.
245. Krausa, M., Doll, J., Schorb, K., Boke, W. an. Hambitzer, G. Fast electrochemical detection of nitro- and aminoaromates in soils and liquids. *Propellants, Explosives, Pyrotechnics*, 1997, **22**, 156–159.
246. Wang, J., Bhada, R. K., Lu, J. and McDonald, D. Remote electrochemical sensor for monitoring TNT in natural waters, *Analytica Chimica Acta*, 1998, **361**, 85–91.
247. Wang, J., Lu, F., MacDonald, D., Lu, J., Ozsoz, M. E. S. and Rogers, K. R. Screen-printed voltammetric sensor for TNT. *Talanta*, 1998, **46**, 1405–1412.
248. Wu, D., Singh, J. P., Yueh, F. Y. and Monts, D. L. 2,4,6-Trinitrotoluene detection by laser-photofragmentation-laser-induced fluorescence. *Applied Optics*, 1996, **35**, 3998–4003.
249. Boudreaux, G. M., Miller, T. S., Kunefke, A. J., Singh, J. P., Yueh, F.-Y. and Monts, D. L. Toward development of a photofragmentation-laser induced fluorescence (PF-LIF) laser sensor for detection of 2,4,6-trinitrotoluene (TNT) in soil and groundwater. *Applied Optics*, 1999, **38**, 1411–1417.
250. Rouhi, A. M. Land mines: horrors begging for solutions, *Chemical & Engineering News*, 1997, March 10, 14–22.
251. Trevelyan, J. *Mines and Minefields. Defining the Problem.* University of Western Australia, Department of Mechanical and Materials Engineering, Nedlands 6907, Western Australia, 1997.
252. Anon. *Landmines: Time for Action.* Report No. 0574/002, International Committee of the Red Cross, Geneva, 1995.
253. Anon. *Anti-personnel Mines: An Overview.* International Committee of the Red Cross, Geneva, 1997.
254. Tsipis, K. *Landmine Brainstorming Workshop.* Report No. 27, Massachusetts Institute of Technology, Cambridge. MA, 1996.
255. Antonic, D. and Ratkovic, I. *Ground Probing Sensor for Automated Mine Detection.* Ministry of Interior, Department of Small Arms, Special Equipment and Security, HR-10000 Zagreb, Croatia, 1998.
256. Trang, A. H. and Czipott, P. V. Characterization of small metallic objects and nonmetallic antipersonnel mines. *Conference on Detection and Remediation Technologies for Mines and Minelike Targets II*, Orlando, FL, 1997. *SPIE Proceedings*, 1997, **3079**, 372–383.
257. Young, R. and Helms, L. *Applied geophysics and the detection of buried munitions.* http://w2.hnd.usace.army.mil/oew/tech/rogpprl.html. (Access authorization required.) US Army Engineering and Support Center, Huntsville, AL, 1998.
258. Bruschini, C. and Gros, B. A survey of current sensor technology research for the detection of landmines. *International Workshop on Sustainable humanitarian demining (SusDem'97)*, Zagreb, Croatia, 1997, pp. 6.18–6.27.
259. Guerne, F., Gros, B., Schreiber, M. and Nicoud, J.-D., Detec-1 and Detec-2: GPR mine sensors for data acquisition in the field. *International Workshop on Sustainable Humanitarian Demining (SusDem'97)*, Zagreb, Croatia, 1997, pp. 5.34–5.39.
260. Anon. *Stand-off mine detection radar system.* http://www.jaycor.com/eme/smdr.htm. Technical product information, Jaycor, San Diego, CA 92121, 1998.
261. Scheerer, K. Airborne multisensor system for the autonomous detection of landmines. *EUREL International Conference on Detection of Abandoned Land Mines*, Edinburgh, UK, 1996, pp. 183–187.

262. Scheerer, K. Airborne multisensor system for the autonomous detection of landmines. *Conference on Detection and Remediation Technologies for Mines and Minelike Targets II*, Orlando, FL, 1997. *SPIE Proceedings*, 1997, **3079**, 478–486.

263. Wehlburg, J. C., Keshavmurthy, S. P., Dugan, E. T. and Jacobs, A. M. Geometric considerations relating to lateral migration radiography (LMR) as applied to the detection of land mines. *Conference on Detection and Remediation Technologies for Mines and Minelike Targets II*, Orlando, FL, 1997. *SPIE Proceedings*, 1997, **3079**, 384–393.

264. Anon. *Vehicle mounted mine detector*. http://www.demining.brtrc.com/solut/ svhcmmd.htm. US Army CECOM NVESD, Fort Belvoir, VA, 1997.

265. McFee, J., Aitken, V., Chesney, R., Das, Y. and Russell, K. A multisensor, vehicle-mounted, teleoperated mine detector with data fusion. *Conference on Detection and Remediation Technologies for Mines and Minelike Targets III*, Orlando, FL, 1998. *SPIE Proceedings*, 1998, **3392**, 1082–1093.

266. Khanna, S. M., Paquet, F., Apps, R. and Seregelyi, J. S. New hybrid remote sensing method using HPM illumination/IR detection for mine detection. *Conference on Detection and Remediation Technologies for Mines and Minelike Targets III*, Orlando, FL, 1998. *SPIE Proceedings*, 1998, **3392**, 1111–1121.

267. Miller, C. Labs developing means to sniff out mines chemically and electronically. *Sandia Lab News*, 1997.

268. Phelan, J. M. and Webb, S. W. *Environmental fate and transport of chemical signatures from buried landmines-Screening model formulation and initial simulations*. Report No. SAND97-1426, Sandia National Laboratories, Albuquerque, NM, 1997.

269. Phelan, J. M. and Webb, S. W. Chemical detection of buried landmines. *Third International Symposium on Technology and the Mine Problem*, 1998.

270. Phelan, J. M. and Webb, S. W. Simulation of the environmental fate and transport of chemical signatures from buried landmines. *Conference on Detection and Remediation Technologies for Mines and Minelike Targets III*, Orlando, FL, 1998. *SPIE Proceedings*, 1998, **3392**, 509–520.

271. Anderson, D. M., Kistner, F. B. and Schwarz, M. J. *The mass spectra of volatile constituents in military explosives*. Special Report No. 105, US Army Cold Regions Research and Engineering Laboratory, Hanover, NH, 1969.

272. Murrmann, R. P., Jenkins, T. F. and Leggett, D. C. *Composition and mass spectra of impurities in military grade TNT vapor*. Special Report No. 158, US Army Cold Regions Research and Engineering Laboratory, Hanover, NH, 1971.

273. Jenkins, T. F., Murrmann, R. P. and Leggett, D. C. Mass spectra of isomers of trinitrotoluene. *Journal of Chemical and Engineering Data*, 1973, **18**, 438–439.

274. Jenkins, T. F., O'Reilly, W. F., Murrmann, R. P., Leggett, D. C. and Collins, C. I. *Analysis of vapors emitted from military mines*. Special Report No. 193, US Army Cold Regions Research and Engineering Laboratory, Hanover, NH, 1973.

275. Jenkins, T. F., O'Reilly, W. F., Murrmann, R. P. and Collins, C. I. *Detection of cyclohexanone in the atmosphere above emplaced antitank mines*. Report No. 203, US Army Cold Regions Research and Engineering Laboratory, Hanover, NH, 1974.

276. Johnston, J. M., Williams, M., Waggoner, L. P., Edge, C. C., Dugan, R. E. and Hallowell, S. F. Canine detection odor signatures for mine-related explosives. *Conference on Detection and Remediation Technologies for Mines and Minelike Targets III*, Orlando, FL, 1998. *SPIE Proceedings*, 1998, **3392**, 490–501.

277. Lewis, N. S., Lonergan, M. C., Severin, E. J., Doleman, B. J. and Grubbs, R. H. Array-based vapor sensing using chemically sensitive carbon black-polymer resistors. *Detection and Remediation Technologies for Mines and Minelike Targets II*, Orlando, FL, 1997. *SPIE Proceedings*, 1997, **3079**, 660–670.

278. Doleman, B. J., Sanner, R. D., Severin, E. J., Grubbs, R. H. and Lewis, N. S. Use of compatible polymer blends to fabricate arrays of carbon black-polymer composite vapor detectors. *Analytical Chemistry*, 1998, **70**, 2560–2564.

279. Dickinson, T. A., White, J., Kauer, J. S. and Walt, D. R. A chemical-detecting system based on a cross-reactive optical sensor array, *Nature*, 1996, **382**, 697–700.

280. White, J., Kauer, J. S., Dickinson, T. A. and Walt, D. R. Rapid analyte recognition in a device based on optical sensors and the olfactory system. *Analytical Chemistry*, 1996, **68**, 2191–2202.

281. Michael, K. L., Taylor, L. C., Schultz, S. L. and Walt, D. R. Randomly ordered addressable high-density optical sensor arrays. *Analytical Chemistry*, 1998, **70**, 1242–1248.

282. Albert, K. J., Dickinson, T. A., Walt, D. R., White, J. and Kauer, J. S. Designing optical sensor arrays with enhanced sensitivity for explosives detection. *Conference on Detection and Remediation Technologies for Mines and Minelike Targets III*, Orlando, FL, 1998. *SPIE Proceedings*, 1998, **3392**, 426–431.

283. Yang, J.-S. and Swager, T. M. Porous shape persistent fluorescent polymer films: An approach to TNT sensory materials. *Journal of the American Chemical Society*, 1998, **120**, 5321–5322.

284. McGill, R. A., Abraham, M. H. and Grate, J. W. Choosing polymer coatings for chemical sensors. *Chemtech*, 1994, **24**, 27–37.

285. McGill, R. A., Mlsna, T. E., Chung, R., Nguyen, V. K., Stepnowski, J., Abraham, M. H. and Kobrin, P. Sorbent coatings for nitroaromatic vapors: Applications with chemical sensors. *Conference on Detection and Remediation Technologies for Mines and Minelike Targets III*, Orlando, FL, 1998. *SPIE Proceedings*, 1998, **3392**, 384–389.

286. Kobrin, P., Seabury, C., Linnen, C., Harker, A., Chung, R., McGill, R. A. and Matthews, P. Thin film resonators for TNT vapor detection. *Conference on Detection and Remediation Technologies for Mines and Minelike Targets III*, Orlando, FL, 1998. *SPIE Proceedings*, 1998, **3392**, 418–423.

287. Fliermans, C. B. and Lopez-de-Victoria, G. Microbial mine detection system (MMDS). *Conference on Detection and Remediation Technologies for Mines and Minelike Targets III*, Orlando, FL, 1998. *SPIE Proceedings*, 1998, **3392**, 462–468.

288. Fisher, M., Cumming, C., la Grone, M. and Taylor, R. An electrostatic particle sampler and chemical sensor system for landmine detection by chemical signature. *Conference on Detection and Remediation Technologies for Mines and Minelike Targets III*, Orlando, FL, 1998. *SPIE Proceedings*, 1998, **3392**, 565–574.

289. Chambers, W. B., Rodacy, P. J., Jones, E. E., Gomez, B. J. and Woodfin, R. L. Chemical sensing system for classification of mine-like objects by explosives detection. *Conference on Detection and Remediation Technologies for Mines and Minelike Targets III*, Orlando, FL, 1998. *SPIE Proceedings*, 1998, **3392**, pp. 453–461.

290. Desilets, S., Haley, L. V. and Thekkadath, U. Trace explosives detection for finding landmines. *Conference on Detection and Remediation Technologies for Mines and Minelike Targets III*, Orlando, FL, 1998. *SPIE Proceedings*, 1998, **3392**, 441–452.

291. Fair, R. B., Pollack, M. and Pamula, V. MEMS devices for detecting the presence of explosive material residues in mine fields. *Conference on Detection and Remediation Technologies for Mines and Minelike Targets III*, Orlando, FL, 1998. *SPIE Proceedings*, 1998, **3392**, 409–417.

292. McFee, J., Cousins, T., Jones, T., Brisson, J. R., Jamieson, T., Waller, E., LeMay, F., Ing, H., Clifford, E. and Selkirk, B. A thermal neutron activation system for confirmatory non-metallic land mine detection. *Conference on*

Detection and Remediation Technologies for Mines and Minelike Targets III, Orlando, FL, 1998. *SPIE Proceedings*, 1998, **3392**, 553–564.

293. Hibbs, A. D., Barrall, G. A., Czipott, P. V., Lathrop, D. K., Lee, Y. K., Magnuson, E. E., Matthews, R. and Vierkotter, S. A. Landmine detection by nuclear quadrupole resonance. *Conference on Detection and Remediation Technologies for Mines and Minelike Targets III*, Orlando, FL, 1998. *SPIE Proceedings*, 1998, **3392**, 522–532.

294. Anon. *Laser ordnance neutralization*. http://www.zeus.sparta.com/overview.html, Sparta, Inc., Huntsville, AL 35805, 1998.

295. Anon. *Humanitarian Demining Equipment Catalog*. http://www.demining. brtrc.com/catalog/catindex.htm, US Army CECOM NVESD, Fort Belvoir, VA, 1997.

296. Rosen, E. M. and Altshuler, T. W. Clutter and target signature statistics from the DARPA background clutter experiment. *Conference on Detection and Remediation Technologies for Mines and Minelike Targets III*, Orlando, FL, 1998. *SPIE Proceedings*, 1998, **3392**, 1054–1070.

297. George, V., Altshuler, T. W. and Rosen, E. M. DARPA background clutter data collection experiment excavation results. *Conference on Detection and Remediation Technologies for Mines and Minelike Targets III*, Orlando, FL, 1998. *SPIE Proceedings*, 1998, **3392**, 1000–1011.

298. Andrews, A. M., George, V. and Altshuler, T. W. Quantifying performance of mine detectors with fewer than 10,000 targets. *Conference on Detection and Remediation Technologies for Mines and Minelike Targets II*, Orlando, FL, 1997. *SPIE Proceedings*, 1997, **3079**, 273–280.

Index